Reprints of Economic Classics

THE LABORING CLASSES OF ENGLAND

THE LABORING CLASSES OF ENGLAND

ESPECIALLY THOSE CONCERNED IN AGRICULTURE AND MANUFACTURES

IN A SERIES OF LETTERS

BY AN ENGLISHMAN

[WILLIAM DODD]

ALSO

A Voice from the Factories

A POEM IN SERIOUS VERSE

[CAROLINE E. S. NORTON]

[1848]

AUGUSTUS M. KELLEY • PUBLISHERS
FAIRFIELD 1976

Second American Edition 1848

(*Boston*: *Published by* John Putnam, *81 Cornhill, and sold by* The Author, *11 Mount Vernon Avenue*, 1848)

Reprinted 1976 by

AUGUSTUS M. KELLEY • PUBLISHERS

Fairfield, New Jersey 07006

Library of Congress Cataloging in Publication Data

[Dodd, William] b. 1804 (June 18)
The laboring classes of England.

(Reprints of economic classics)
Reprint of the 2d ed., 1848.
A voice from the factories, by C. E. S. Norton: p.
1. Labor and laboring classes--Great Britain.
2. Factory system--Great Britain. 3. Agricultural laborers--Great Britain. 4. Machinery in industry.
I. Norton, Caroline (Sheridan) 1808-1877. A voice from the factories. 1973. II. An Englishman.
III. Title.
H8389.D73 1973 301.44'42'0942 68-55703
ISBN 0-678-00961-9

PRINTED IN THE UNITED STATES OF AMERICA
by SENTRY PRESS, NEW YORK, N. Y. 10013
Bound by A. HOROWITZ & SON, FAIRFIELD, N. J.

THE LABORING CLASSES OF ENGLAND,

ESPECIALLY THOSE CONCERNED IN

AGRICULTURE AND MANUFACTURES;

IN A SERIES OF LETTERS.

By an Englishman.

ALSO,

A Voice from the Factories,

A POEM, IN SERIOUS VERSE.

SECOND EDITION.

BOSTON:
PUBLISHED BY JOHN PUTNAM, 81 CORNHILL.
AND SOLD BY THE AUTHOR,
11 Mount Vernon Avenue.
1848.

CONTENTS.

THE

LABORING CLASSES OF ENGLAND

LETTER I.

INTRODUCTION.

In offering the following work to the public, I have been actuated by a desire to diffuse as widely as possible the information it contains; believing it will be interesting and instructive to every well-wisher of the human race.

I have been led to publish the following facts, in consequence of the curiosity manifested by almost every person with whom I have become acquainted in America, to know my history, &c. So great has been the desire to question me upon this subject, that I have felt it, sometimes, to be my duty to refuse to give any information; my own feelings requiring me *to forget, as far as possible*, the injuries of former years. Whenever I feel my heart beat quicker, occasioned by a retrospective view of my sufferings, my peace of mind demands that I should instantly cry, "peace, be still." I believe, that had I fallen from some distant planet in the Solar system, the desire to know my history, and that of my species, could not have been greater. A single glance at my person, as I walk along the street, or stand in the presence of any one, is sufficient to awaken this curiosity in a country like America, where no such cripples are made by hard labor; but in England, where they are to be met in almost every street, it is very different. In order, therefore, to gratify this laudable curiosity, and spare my own feelings, publishing became necessary.

It may be asked how I gained the whole of my information upon this subject. To this I would answer, my situation has been in many respects peculiar. For twenty-five years of my short life, I have been actively engaged as an operative in the English factories. I am not aware that any one else who has published upon the factory system can make a similar assertion. I have not only toiled, but have been a sufferer from protracted mill labor to a painful extent. My experience, therefore, of the factory system has been dear-bought experience. I can speak feelingly, and I trust temperately. I have endeavored to avoid to the uttermost, every unguarded expression, every word which it would not become an humble operative to use; and I can add with truth, that I am not conscious of one unkindly or resentful feeling towards any human being.

In addition to the experience I have had in factories, I was employed in part of the years 1841 and 1842, by a benevolent Nobleman in London to assist him in his laudable endeavors to benefit the laboring classes. It may be interesting to the American reader to know, that my salary under this Nobleman was forty-five shillings per week, (about $11,) and coach hire, while travelling, and twenty shillings per week, ($5,) while stationary in London. Under this engagement I travelled through the West Riding of Yorkshire, Lancashire, Cheshire and Derbyshire; and being well supplied with letters of introduction, I had ample opportunities of conversing with all parties likely to afford me any information on the subject of factory life. In particular, I waited upon Clergymen of various denominations, Manufacturers, Surgeons, Inspectors and Overlookers. I had also opportunities of studying the habits and manners of the operatives, in the mills, cottages, places of amusement, public houses, &c., and of investigating the various causes of decrepitude, mutilation or death;—whether arising from long hours of labor, or accidents by machinery.

The facts contained in this volume have been carefully inquired into on the spot, and in many cases taken from the parties themselves, and corroborated by others not interested in the matter. I have no doubt the reader will be interested in perusing the following letters, which, with many others, I received from this nobleman while in his service.

[No. 1.]

Oct. 12, 1841.

DEAR ——. You have discharged your *commission* admirably, and I am much obliged to you for the trouble you take, and the accuracy with which you furnish details.

I trust you will derive from your present duty that real satisfaction, which is the portion of those who labor, in God's name, for the welfare of their fellow creatures. I commit you most heartily to His care, and wish you every happiness in this world, and in that which is to come.

Faithfully yours, A——.

[No. 2.]

St. G—— House, Nov. 24, 1841.

DEAR ——. So far from thinking that you travel beyond your duty, when you write to me your opinions on all matters affecting the moral condition of the working classes, I am exceedingly pleased with your remarks; I altogether concur in them, and request you to continue your observations. I have always been convinced that a reduction in the hours of labor is only a preliminary to the measures we must introduce for the benefit of the working classes; but it is an *indispensable* preliminary. We must first settle this just principle, and then go on, by God's blessing, to draw long advantages from it. Limited as I am *in* Parliament, and *out* of it, I cannot undertake more than one thing at a time; but I think of a great many, and hope to be able hereafter to effect a few of them.

Your labors have been very serviceable. It must be a pleasure to you, to find yourself by God's mercy, in a way to be of use to your fellow sufferers, to make at least an ingenuous effect. I hope that your remaining days may be so assured to you in comfort, that you may have leisure and means to pursue your *plans* for the welfare of the operatives.

Your faithful servant, A——.

[No. 3.]

March 31, 1842.

——. Pray go to the house of Mrs. Torvey, 41 A——t Street, Regent's Park.

You will there see a poor girl whose arm has been torn off by a wheel in a silk mill. Pray talk to her, and tell me what you think of the case.

You will be able to judge whether I can assist her by giving her a *false hand*, such as you have.

Your humble servant, A——.

My "plans" alluded to in letter No. 2, were chiefly the establishment of a self-acting asylum in the neighborhood of London, for the reception of the thousands of destitute factory cripples, in which they might be provided with the means of spending the remainder of their days in comfort, and in preparing for another and a better world. I had also formed some plans for preventing, as far as human means could prevent, the making of cripples in future. Although I did not succeed in carrying out these desirable objects, it was gratifying for me to know that "I had discharged my *commission* admirably," and that my "labors had been very serviceable."

My statements respecting agricultural laborers, have been chiefly derived from the reports of Commissioners laid before Parliament; and which were borne out by my own observation and experience.

In a country like America, where all men, in the eye of the law, are born equal, it is extremely difficult for the majority of readers to comprehend the real position of the laboring classes, in countries under a monarchical form of government. It is, in the first place, difficult to understand what is meant by "classes." For the information of such readers, it may be proper to say a few words upon this subject.

English society may be conveniently divided into eight classes:

1st. *The Royal Family.*—Under this general term are comprehended all who are of the blood royal.

2d. *The Nobility.*—In this class we have Archbishops, Dukes, Marquisses, Earls, Viscounts, Bishops, Barons, &c. They are commonly denominated "the upper ten thousand."

3d. *The Millionaires*, commonly called "the vulgar rich." This class comprehends a great number of individuals who have amassed immense wealth by manufactures, commerce, railroad speculations, &c.

4th. This class is composed of the clergy, professional gentlemen, merchants, tradesmen, &c. The gentlemen composing this class, with the exception of the humbler order of the clergy of all denominations, are well remunerated for their services, perhaps better than a similar class in any other country on the globe.

5th. The higher order of Mechanics, known as "skilled laborers," (from their being obliged to pay large fees, and to serve an apprenticeship of seven years to the trade which they follow,) shopkeepers, &c., compose this class. Generally speaking, they are an industrious and intelligent class, and are sufficiently remunerated for their services to enable them to bring up their families in a respectable manner, and to lay by something for the comforts of old age.

6th. This class comprehends a great number of individuals who get their living by the "sweat of their brow," but who are not required to serve seven years at their trade or calling. Manufacturing, agricultural, and many other kinds of laborers, come under this head. This class is a hard-working, ill-paid, and ill-used set of human beings; frequently dying with every symptom of premature decay, at from 35 to 50 years of age.

Each individual is compelled to pay taxes to the government, the taxes being levied upon their provisions, clothes, furniture, &c. They are also compelled to obey upwards of 1500 laws, without having a voice in making or amending one. Their appeals to Parliament by petition, are scarcely ever listened to, unless seconded by some of the "privileged" classes. It is to this class my observations in this work principally apply.

7th. *Paupers.* Of this class there is known to be in Great Britain and Ireland, 4,000,000 of individuals, of all ages and both sexes. It may be said of them, that they have lost all but their *integrity*, and that there is little hope left for them, of bettering their condition in this world.

8th. This is a class who have lost what the class above still retain, their honor, integrity, good names; who have no recognized means of existence, but live by their wits upon the property of others. Thieves, gamblers, prostitutes, and the like, are of this class.

The outlines of these several classes are broad and well defined; there are, however some peculiarities common to two or more classes. Thus if we couple together classes 1 and 2, we shall have a mass of individuals commonly known as the "head;" and following the same rule with 7 and 8, we get what is called the "tail" of society.

The first four may be called "privileged classes;" and the last four *non*-privileged classes. The first five as *law-making classes*, the last three as classes having nothing to do with the laws but to obey, to do, and to suffer as others may direct.

It is a matter of some doubt with the writer, whether there are to be found in the world the same number of people enjoying equal privileges, as the first four classes; or a portion of any community enduring privations and sufferings such as are patiently endured by the last four.

There is something which makes some of these classes *attract* and *repel* each other. Thus the poorer portion of class 2, have a great affinity for class 3, and many of class 3 having got all but a "title," reciprocate this sympathy, and marriage is the consequence. Repulsion takes place when any members of the "head" are brought into contact with a member of the "tail."

The *ascent* in these classes is attended with difficulty and danger to the adventurous individual who attempts it; the *descent* is accomplished much easier.

I am a native of class 4, and was reduced in childhood to class 6. I rose again after I had quitted the factories to my native element 4; after I had lost my arm I again sunk to 6, and it was with great difficulty I prevented myself falling to 7.

With respect to the beautiful poem, I may say, that I do not know the name of the author. It appears, it was printed for private circulation among the upper classes;

a copy of it was put into my hand by the late celebrated publisher, Mr. Murray, of Albemarle Street, London.

In perusing the following pages, it will be necessary that the reader should bear in mind, that the author is a working man, that he never went to school, that he is here describing things which he has witnessed in every-day life, and that his observations are confined to that portion of society in which he has lived and moved.

With these preliminary remarks, I leave the work to the candid reader, and to God's blessing, believing that it does not contain a single sentence which on my death-bed I could wish to erase.

THE AUTHOR.

P. S. Should any lady or gentleman feel desirous of seeing for themselves the horrors of the English factory system, as it is stamped on my person, a letter to my address, post paid, will be attended to.

No. 8 *Mount Vernon Avenue.*

LETTER II.

INCIDENTS IN THE LIFE OF THE AUTHOR.

In the early part of the present century, the author's mother, who we shall call the widow Graham, had been left to struggle with the ills of life, and to feel, together with her offspring, the painful realities of want and suffering; but too frequently the lot of the widow and the fatherless.

For a time she could not well understand the position in which she was placed; but the pressing calls of her family, gradually brought her to a sense of her real condition. Being descended from an ancient Scottish family, and being still strong and active, she was not the person to shrink from difficulties. It is true the task of bringing up four children, the oldest of whom was but 9 and the youngest 4 years of age, by her own endeavors, was no easy one. It is also true there was a way by which she could be lightened of her burden, viz: by placing her children in the workhouse; but this was repugnant to her feelings. Although she had seen better days, it must not be supposed that this repugnance arose from any family pride, or from any improper feelings with respect to her situation. She was duly resigned to her condition, and fully determined to discharge her duty to the utmost of her power.

To those persons who are acquainted with the manner in which the parish children in England were treated in the early part of the present century, how they were drafted off by boat loads into the factory, it will not be surprising that she refused to listen to all advice from friends to seek relief in that way. It was in vain they represented the impossibility of bringing up her family without assistance; she had formed her plans not to part from her children under any circumstances.

About this time the large manufacturers of Lancashire, and Yorkshire, having nearly worked up all the parish children from London, Birmingham, and other large towns, sent their agents into the small towns and villages around to pick up any poor families whom they might meet with. An engagement with these agents would at once have relieved widow Graham of three of the children; but this was nearly the same thing as sending them to the work-house, and like the other was abandoned.

What was to be done? Work the children must, at something; for one pair of hands, however industriously employed, could not maintain five persons, besides paying house rent, fire and taxes. Accordingly she agreed to send the two oldest girls to a factory in the neighborhood on trial, then her favorite little boy Jemmy, and lastly her youngest girl.

They all continued to work in factories many years; the result was the oldest girl met an untimely death, the youngest was taken away to save her from the same fate, and the two others became cripples for life.

But it is to the boy Jemmy that the reader's attention is called.

When first he was sent to the factories, being but 5 years and 9 months old, he was too short to reach the top of the frame at which he was set to work, and a block of wood was given him to stand on, in order that he might be enabled to get to his work properly. The hours which he was obliged to work were from 6 o'clock in the morning till half past 7 in the evening; with one hour and a half for meals, with 12 working hours for 5 days, and 9 on Saturday. For this employment he received 24 cents the first week, and 36 cents the second, at which rate he continued for several months, when his wages were advanced to 48 cents per week.

The little fellow could not at this early period of his life be supposed to be worth much as a laborer, and probably the small amount here mentioned was the full value of his services; be this as it may, the punishment to him arising from standing so many hours without being permitted to sit down was very severe, and ought never to be required of children for such a pittance, or in short, under any circumstances. He continued to increase in his qualifications and was several times advanced, till at fourteen years of age, having then been 8 years in the factories, he was capable of earning 72 cents per week, which was a little more than the average for children of his age. During these 8 years he went through a series of uninterrupted, unmitigated suffering, such as very rarely falls to the lot of mortals so early in life, except to those situated as he was, and such as he could not have endured had he not been strong and of a good constitution.

At the age of 8 or 9, his limbs began to show symptoms of giving way, under the excessive fatigue to which he was subjected. He constantly complained of weariness, pains in the knees and ancles, and was ever ready to sit himself down in the factory, on the road, or in almost any place, whenever and wherever an opportunity presented itself, even for half a minute.

Every precaution was taken that the humble means of his widowed mother would permit, to prevent her favorite, her only boy, from being made a cripple; but in vain. Oils, flannel bandages, strengthening plasters and mixtures, were incessantly applied; and every thing but the right one, (viz. taking him from the work,) were one by one tried, rejected, and abandoned. In defiance of all these remedies, he became from excessive labor, a confirmed cripple for life. His knees gave way and gradually sunk inwards till they touched each other, thus forming a kind of arch for the support of the body. At 12 years of age the easiest position in which he could stand, was with his feet about 10 or 12 inches apart, his knees resting as above, with the centre of gravity crossing the thigh and leg bones and falling within the feet.

The school in which he was thus placed was any thing but favorable to a life of morality. Under the same roof were more than 100 children and young persons of both sexes, going together in the morning, associating with each other through the day, returning again in the evening, with no moral restraint upon their actions, no example set them worthy of imitation. On the contrary, low, vulgar, brutal language, swearing, singing immoral songs, and acts of gross indecency, were not only tolerated, but in many instances actually countenanced and encouraged. A factory conducted thus was not a very desirable place to train up a child in, and many a time did it grieve the heart of his mother to hear him, in answer to her inquiries as to how he had come by a bruise or cut on the head or back, tell how he had been beat by the overlooker or spinner, and how he swore he would kill him if he did not work faster. Anxiously did she inquire of her friends for some more suitable employment for her boy; but on account of his deformity, which had become now quite conspicuous, none could be found.

The situation of Jemmy at 14 was truly distressing. He could not associate with the other boys at play in his leisure moments; neither could he go to the Sunday school, as he had done in his younger days; on the contrary, he sought every opportunity to rest himself, and to shrink into any corner to screen himself from the prying eyes of the curious and scornful! During the week days he frequently counted the clock, and calculated how many hours he had still to remain at work. His evenings were spent in rubbing his joints and preparing for the following day; after which he retired to cry himself to sleep, and pray that the Lord would release him from his sufferings before morning.

On finding himself settled in the factories, as it was then pretty evident he could get no other employment, he began to think of getting a little higher in the work, and speaking to the master upon the subject, he got advanced to a place where the labor was not so distressing, but where the care and responsibility were greater.

Soon after his advancement to this place, Jimmy met with an accident which had very near been fatal. By some means his coat got entangled in the straps of the machinery, and finding himself lifted from the floor, with a prospect of being dashed against the floor above, he gave a sudden jerk, and the coat being old and saturated with oil, broke away, and thus saved his life. A few months after he had another, and still more narrow escape from death. This was partly owing to the want of sufficient precaution in boxing off the machinery.

When about 15 years of age, a circumstance occurred to him, which does not often fall to the lot of factory children, and which had a great influence on his future life. He happened one day to find an old board lying useless in a corner of the factory. On this board, with a piece of chalk, he was scrawling out as well as he was able, the initials of his name, instead of attending to his work. Having finished the letters, he was laying down the board and turning to his work, when to his great surprise, he perceived one of his masters looking over his shoulder. Of course, he expected a severe scolding; but the half smile upon his master's countenance suddenly dispelled his fears. This gentleman, who was a member of the society of Friends, kindly asked Jimmy several questions about reading and writing, and being informed that the two letters on the board before him, contained the sum total of all the knowledge the boy possessed in these matters, he kindly gave him 2 pence (4 cents,) to purchase a slate and pencil, pens, ink and paper. This sum of money he continued to allow the boy weekly for several years, always inspecting his humble endeavors, and suggesting any improvements which he thought necessary.

Thus an opportunity was afforded him, which, with a few presents of books, was the means, under Providence, of laying the foundation of a tolerable education, for a working man. This kindness on the part of his master is still fresh in his memory. He speaks of it as one of the bright spots in his checkered life.

With this encouragement, and impelled by the activity of his own mind, and an irresistible thirst after knowledge, he set himself earnestly to the acquisition of such branches of education as he thought might better his condition in after life; and although he had still his work to attend, he soon found himself in possession of a tolerable share of mathematics, geography, history, &c.

Now that he began to derive pleasure from the perusal of books, (and in fact, it was the only source of pleasure he had) he did not omit any opportunity of gratifying this desire, but particularly on the Sabbath day. His usual custom in the summer months was to take a book in one pocket, and a crust of bread in another, and thus provided with food for the mind and body, go forth on a Sunday morning to a retired and secluded wood, about 2 miles from the town in which he lived, and there spend the day alone. On the banks of a rivulet which ran through the wood, he has sat for hours absorbed in study, unperceived by mortal eye, with nothing to disturb the solitude of the place but the numerous little songsters that kept up a continual concert, as if to make it more enchanting to his imagination.

These visits to his summer retreat he speaks of as seasons of real pleasure; they were also attended with some advantages in point of health. For a number of years he had enjoyed but a delicate state of health, owing to constant confinement, the smells of the factories, &c.; but these Sunday excursions got him a better appetite for his victuals, and he became more healthy and strong.

He also derived considerable pleasure and improvement from the study of nature, in watching the habits of birds, bees, ants, butterflies, and any natural curiosity that came in his way; and when the evening began to close in around him, and compelled him to return to the habitations of men, he felt a reluctance to leave his quiet and solitary retreat.

On some occasions, when returning from his retreat in the woods on a Sunday evening, he has stood upon an eminence at a distance, and watched the gaily-attired inhabitants taking their evening walk in the fields and meadows around the town, and could not help contrasting their situation with his. They were happy in themselves, anxious to see and be seen, and deriving pleasure from mutual friendship and intercourse; he, with the seeds of misery implanted in his frame, surrounded by circumstances calculated to make him truly unhappy, shrinking from the face of men to a lovely wood, to brood over his sorrows in secret and in silence. They were enjoying the fruits of their industry; but the reward for his, was misery, wretchedness and disease.

So great was the love of books in this youth, that he seized upon all within the circle of his acquaintance, no matter upon what subject, with avidity. On one occasion he was tempted to have recourse to a little of what the world calls policy, in order to gratify his appetite for reading, but which he knew to be wrong.

On Saturdays, the mill usually stopped working at 5 o'clock; then after cleaning himself, he had a few hours to call his own, which were generally spent in his favorite amusement. One fine Saturday evening in June, having provided himself with a book from a circulating library, he took a walk to the ruins of an old castle, a short distance from the town, which had often been to him an agreeable retreat from the noise and bustle of the factory. For the loan of this book he had paid two pence, the sum his master allowed him weekly over and above his wages, and he had got it snugly in his pocket, calculating on the pleasure it would afford him during the week. It chanced, however, to be one of those thinly printed volumes with large margins, and seating himself on the above mentioned ruins, he did not rise till he had finished it. When he rose from his seat the evening was closing in around, and the bats and owls were on the wing; but he had read his book, had exhausted his whole week's stock of amusement. What was to be done? To obtain another volume in the usual way was impossible, he had not another penny in the world; and to be without a book for a whole week seemed very hard. In this dilemma he hit upon a plan, which after a little hesitation he carried into effect. He took the volume back to the librarian and requested him to change it, telling him it did not suit. His request was complied with, and he was thus furnished with amusement for the week.

When about the age of 17, he became acquainted with a young student who was very kind in lending him books, and explaining any difficulty he might be laboring under in his studies. This student also first directed his attention to higher and nobler objects, got our youth to relinquish in part, his Sunday excursions, and go with him occasionally to church. He speaks of these kindnesses as having been of great service to him, and recollects them with gratitude.

To turn his thoughts from his pitiful situation, he attended lectures on various subjects, repeated the simple experiments at home, made some curious models and drawings of machines, and could thus contrive to pass away his leisure time pleasantly. But in proportion as the truths of science were unfolded to his wondering sight, and the mists of ignorance chased from his mind, new desires sprung up which were before unknown, and from being entirely out of his reach, made him occasionally fretful and unhappy.

Being desirous of turning his newly acquired learning to some account, he engaged to keep the books of a tailor, draw out his bills, &c. in the evenings after his labor in the factories was over, by which he earned part of his clothing, and also got an insight into the trade, which was of service afterwards.

From the time our young friend had been put into the factories, he had gradually, but slowly advanced from one process to another, till by the time he had arrived at the age of 20, when he had been in constant practice more than 14 years, he found himself to be a person of some consequence. He was then well acquainted with the various processes of manufacturing woollen cloth, and would have had no hesitation to undertake to make a piece of cloth throughout himself. Besides, his literary knowledge, if I may be allowed the expression, enabled him to undertake to keep the books of the factory, which was to him not only an easier situation, but a more profitable one.

In all the advancing stages of his factory life, from a boy standing on a wooden block, to a clerk in his master's counting room, Jimmy had to comply with the evil and pernicious practice of paying *footing;* that is, at every step a person takes in his upward progress, one or two shillings are demanded to be spent in drink, by the work people, to which they contribute a small sum in order to make up a jollification. From these scenes, so contrary to his habits, he was always glad to retire to his books.

On the introduction of some improved machinery for finishing woollen cloth, into the factory, James was appointed to superintend it. This machinery being of a much superior description to any previously in use, it was placed in a room by itself, and all communication with that room shut off, except to the person attending the machine. In this room he worked for several years, being always locked in by himself. One day as he was busily engaged in his occupation, he had his right hand taken into the machine and injured considerably among the bones in the wrist, losing at the same time one finger end. Five years afterwards the right arm had to be amputated, in consequence of some of the bones being injured as is supposed, by this accident, and having still to continue to work.

Although he was not, at this time, constantly employed within the mills, but had to attend to the packing department in the warehouse, and any other place about the works where he might be required, yet still the effects of former years of factory toil were on him, still his life was one of suffering, although not to so great a degree; and he had it now in his power to procure comforts which were before unknown to him, and lived more like a Christian than formerly.

An easy clerk's situation being now vacant, he was advised by some friends to avail himself of the opportunity, and thus free himself totally from the factories, especially as he had several influential friends to forward his views.

He mentioned the subject to his masters, who made such advantageous offers as induced him to remain with them.

In 1834, the law for the regulation of factories in England was about being put in force. James Graham being then an overlooker, had to take the children to the doctor to be examined, and get certificates that they were of the required age. It was with the utmost difficulty he could persuade the doctor to certify that they were 9 years old, although some of them were in their eleventh year, their stunted, diminutive, and sickly appearance being so much against them.

One of the most trying occurrences in all his factory experience, took place in the following winter. About 8 months previously he had had a youth of about 17 years of age placed under him, for the purpose of learning some of the higher branches of the business. Having one day given this youth directions what to do, and gone up to the room above for the purpose of superintending some other part of the works, he noticed one branch of the machinery suddenly stop. On going to ascertain the cause, he met several persons running towards him, who said, "Tom has got into the Gig and is killed." He ran down in haste, but it was too true; he was strangled. A great many bones were broken, and several ghastly wounds were inflicted on his person.

After his mangled body was extracted from the machinery, by unscrewing and taking the machine in pieces, it was laid in a recess in the ground floor, the same in which the accident occurred, to await a coroner's inquest, the works being all stopped and the hands dismissed. The reader may imagine the feelings of Graham, as he paced backwards and forwards with folded arms and downcast eyes. It was a cold winter's evening. He had a flickering light burning beside him. Not a sound broke upon the ear, except the wind and rain without, and the water trickling through the wheels within; while the mangled remains of that youth whom he had instructed in his business, and looked upon almost like a son, lay bleeding beside him.

A little while before this, Graham's sister had met with an accident, whereby she had lost part of her hand, and the remainder was rendered nearly useless.

Graham had now been in the factories about 25 years, and began to feel an earnest desire to quit them. These repeated trials, first his own accident, then his sister's, and afterwards the death of his favorite boy, made him look upon the place with any thing but a favorable eye.

Having previously acquainted his masters with his intentions, he commenced a night school by way of practice for himself, teaching the children of the factory two evenings in the week. In this he was assisted by the young masters. This continued about 12 months.

At the close of the year 1836, he settled his affairs with his masters, and having saved a little money he commenced school-keeping. But on account of there being but few working people able to send their children to school, and people in a higher sphere not being willing to send their children to be instructed by one who had never been to school himself, it did not answer his full expectations.

He then tried to get employed in some of the public schools in London, but failed on account of his deformity. For a similar reason he has partly failed in several other things he tried. In 1839, being then 33 years of age, he bound himself an apprentice for three years to a tailor in London, and in little more than a year the right hand which had been crushed in the machinery, got so bad as to oblige him to give that up also.

The remaining part of his savings was now wanted for the cure of his hand, and after having spent all his money and had several months of painful experience, under some of the most skilful surgeons of London, he was obliged to submit to amputation as the only means of saving his life.

On recovery he set himself to teach the *left* hand the knowledge previously possessed by the *right* hand, such as the use of the pen, needle, sheers, razor, &c. He also invented, with a little assistance from an ingenious machinist of London, an artificial arm, and several instruments, whereby he has been enabled to work at his trade as a tailor.

His health, as a matter of course, has suffered severely, and although he looks healthy, yet those who know him best say his constitution is quite undermined, in consequence of his former hardships and sufferings. He does not calculate on more than three days' good health at one time. On viewing his person one cannot but perceive at a glance, that he was intended by nature for a stout, able bodied man, although he now stands but 5 feet 1 inch high. He calculates he has lost by his deformity 7 inches and a half, which he says can be proved by unerring natural laws.

Such is a brief history of an English factory cripple. It differs but little, except in his literary attainments, from many of the same class of persons, of whom there are at this time, (1846) upwards of 10,000. Many of these are dependent upon their friends and relatives for support.

Petition after petition has been sent into the two houses of Parliament, to the prime minister, and to the Queen, concerning this unfortunate class of British subjects, but without effect. Had they only been *black* instead of *white*, their case would have been taken into consideration long ago. Or if they had been inhabiting any other portion of the Globe, the far-famed English philanthropists would have found them out; but because they are in England itself, under their very eye, their case is unheeded.

There is, however, one comfort, even to this unfortunate class of human beings, viz. that their sufferings will be but of short duration, and their deformities will not be any barrier to another and a happier state of existence.

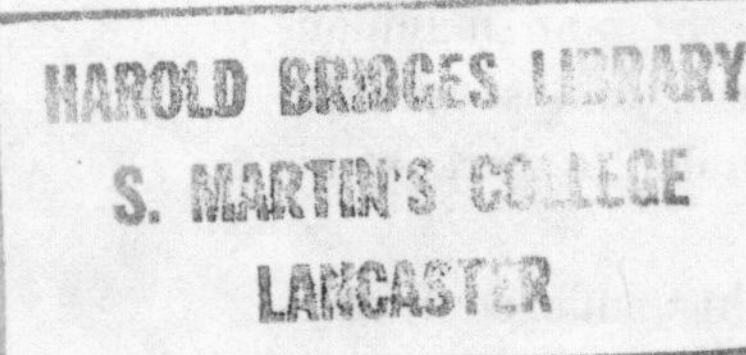

LETTER III.

THE GENERAL CONDITION OF THE LABORING CLASSES.

For the last few years, the attention of the upper and middle class of society in England has been repeatedly and forcibly drawn to the increasing misery and destitution of the laboring classes. In the year 1842, the "Atlas" newspaper proposed to give a prize of £100, £50, and £25, for the 1st, 2d, and 3d best essays respectively, on the causes of, and remedies for the existing distress of the country. From the first of these essays, (which was afterwards made public,) and other sources which may be relied on, I am partly indebted for the following facts.

At former periods of our history we have heard complaints of national distress, and witnessed instances of national decay; but these have been occasioned by causes, and accompanied by symptoms, very different from those which characterize the present phase of social existence in England. For instance, invasion of foreign enemies, loss of national independence, decay of energy and martial spirit, domestic discord, religious persecution, financial embarrassment, sudden changes in the accustomed course of commerce, are all recognized causes and symptoms of the decline of nations. Of none of these do we find a trace in the present condition of England. On the contrary, never, perhaps, was there a period when national prosperity, judged of by these historical tests, stood higher. England stands without dispute, the first naval and commercial power in the world. It would be easy to accumulate facts; but it is not necessary for our present purpose, which is simply to show that the country exhibits, as yet, no decided symptoms of declining wealth, and that whatever may be the evils which afflict society, the want of a sufficient capital to set industry in motion, and to sustain the national burdens, is certainly not among them. Where, then, is the cause of this wide-spread distress?

If neither the political circumstances, the financial conditions, now considered with reference only to the amount of wealth—the economical state of the country, shew any indications of decay and danger, how is it that so many serious men shake their heads with gloomy apprehensions, and at times feel tempted to doubt whether the amount of evil in the present social condition of England does not preponderate over the good. *It is in the condition of the laboring classes that the danger lies.*

Amidst the intoxication of wealth and progress, and the dreams of a millennium of material prosperity to be realized by the inventions of science, the discoveries of political economy, and the unrestricted application of man's energy and intelligence to outward objects, society has been startled by a discovery of the fearful fact, that as wealth increases, poverty and crime increase in a faster ratio, and that in almost exact proportion to the advance of one portion of society in opulence, intelligence, and civilization, has been the retrogression of another and more numerous class towards misery, degradation and barbarism. To speak more specifically, the leading facts to which the evils that, in one shape or other, are continually forcing themselves upon the attention of society, may be reduced, appear to be—1st. The existence of an intolerable mass of *misery*, including in the term both recognized and official pauperism, and the unrecognized destitution that preys, like a consuming ulcer, in the heart of our large cities and densely peopled manufacturing districts. 2d. The condition of a large proportion of the independent laboring class, who are unable to procure a tolerably comfortable and stable subsistence in return for their labor, and are approximating, there is too much reason to fear, towards the gulf of pauperism, in which they will be sooner or later swallowed up, unless something effectual can be done to arrest their downward progress.

With respect to the recognized paupers, it is stated by a writer in Blackwood's Magazine, that in England, Ireland and Scotland, the number is 4,000,000. It is also proved by facts which no one can dispute, that a large proportion of the dense masses of population, crowded together in the lower districts of our large towns, have absolutely no regular and recognized occupations, and live as it were, outlaws upon society. They have, in fact, nothing to look forward to; nothing to fall back upon. One or two facts speak emphatically as to the social deterioration.

In Manchester, in 1839, as many as 42,964 persons, or nearly one sixth of the population, were admitted at different medical charities; and more than one half of the inhabitants are either so destitute or degraded, as to require the assistance of public charity in bringing their offspring into the world. And let it be here remembered that the industrious inhabitants of this large town have done more to uphold what is falsely called the "dignity of the nation," than any other town in the country. In Glasgow, in the five years ending in 1840, as many as 62,051 persons were attacked by typhus fever, a disease generally produced by filth, intoxication and vice. In Liverpool, 35,000 to 40,000 of the lower population live in cellars, without any means of light or ventilation but the door. A like picture is presented to the eye of an attentive observer of society, in Leeds, Birmingham, Brighton, London, and almost all the large towns.

The Journal of Civilization, says, If it were required to draw a strong picture of man, morally and socially degraded by misery, the savage tribes of distant zones would in all probability be selected to sit for it. Yet such darkly shaded originals, such painful realities, need not be sought in remote lands. Let the street beggar or the London thief be followed to his home, (if he have one,) and mankind will be seen existing in degradation as great, enduring misery as sharp, as the South Sea Islanders, or the South Africans in their worst aspect. Amongst them, poverty, vice, ignorance, have no contrast to heighten their effects; but here in England—in London, perhaps at our own back door, wretchedness the most acute, infamy the most shocking, exist upon the same square acre with a high condition of luxury and wealth; and despite their near neighborhood, it may be safely conjectured that the British public know more of the social misery of savage nations, than they do of their own poor. Yet, upon this ignorance, the debased and the criminal are specially legislated for, sometimes incorrectly, always inefficiently.

Amongst the various causes of this state of things, the principle, I believe, is, that of mammon worship. This is one of the vices of modern English society, along with an undue depreciation and neglect of the duties, obligations, and influences of an unseen and spiritual world. The prevalence of this spirit in modern English society, is a fact too obvious to admit of dispute, or to require demonstration.

The very expressions of our common, familiar conversation, testify to it. A "respectable" man has come to signify, a man who lives in a manner which denotes the possession of a certain income; a "successful" man means a man who has succeeded in realizing a certain fortune; a "good match" is synonymous with a marriage to a man of handsome means. The practical working faith of most people for the last century seems to be, that to get on in the world, and realize a certain amount of money and social position, is the *one thing needful.* The sense of duty, which is in itself infinite, has resolved itself into a sort of infinite duty of making money. Our whole duty of man, is, in the first place, to be rich; or, failing in this, in the second place to appear rich. On all hands the doctrine is zealously preached and practised, that "poverty is disgraceful, and that hard cash covers a multitude of sins." Now to the prevalence of this spirit may be directly traced a large portion of the evils of which society complains. This part of the subject might be carried to a much greater length, did our limits allow it; this not being the case, I shall simply draw the attention of the reader to the want of sufficient remuneration for industry, which is one of the principal causes of the evils of the poor.

I find on reference to a book in my possession, that in the time of Henry the VIII, laws were passed relating to food and wages, which placed the working man in a far more favorable position, than he is in England at the present time. The price of provisions, and the wages of labor were settled by act of Parliament. The very same Parliament that passed the law that no corn be exported, also enacted that the rate of wages should be *fourpence* per day; and this circumstance is well worth the attention of the producers of wealth; inasmuch as the above wise and just laws were passed by a House of Commons elected on the principle of universal suffrage. The money of the time of Henry, had a different value to the money in use at the present time; we will therefore see what a day's work was worth in England at the two periods mentioned, viz. 1530 and 1840.

First, we see that any individual employing any other individual, could not, according to act of Parliament, give *less* than FOURPENCE per day. He might give more, but he could not give less.

The price of provisions being regulated by act of Parliament, was as follows, in 1530, to which is added the price in 1840.

	1530.			1840.		
A fat Ox,	£0	16s.	0d.	£20	0s.	0d.
A fat Sheep,	0	1	2	1	15	0
A fat Goose,	0	0	2½	0	3	6
Eggs per dozen,	0	0	0½	0	0	9
Cow,	0	12	0	10	0	0
Fat Pig,	0	3	4	3	3	0
A pair of Chickens,	0	0	1	0	2	0
Wheat per quarter,	0	6	0	3	0	0
Wine per quart,	0	0	1	0	3	4
Table Beer per gallon	0	0	1	0	1	8
Shoes per pair,	0	0	4	0	10	0
	£1	19s.	4d.	£38	19s.	3d.

In 1530, you see there was something like justice meted out to the working man. You will perceive that the laborer, in the course of 20 weeks, could earn as much as would purchase the list of articles enumerated above; but the laborer had a greater advantage than appears at first sight; the act distinctly specifies that the employer *must* give *fourpence per diem*, at the least, so that the laborer was at liberty to hire himself or not; and, you may rest assured, that wages were oftener *above* the fourpence per day, than at it. Again, while the articles above mentioned were brought to market, and could not be sold, only at a certain price, beyond which they dare not be sold, yet the person selling was often compelled to sell them at a lower price. Thus the working man had a double advantage to what he has now; because, although provisions might be *lower* in price, and the rate of wages *higher*, yet wages could not be reduced lower, nor provisions higher, than the act specified.

Now in 1840 the average rate of wages was about ten shillings per week in England. This I believe is admitted by all who have wrote upon the subject. You will thus see by a little calculation that the working man was compelled to work seventy-eight weeks, for the same amount of comforts that he could purchase in 1530 for twenty weeks labor.

It must be, therefore, plainly evident, that the condition of the people, as far as plenty to eat, drink and wear, were concerned, was far preferable to what it is now, because the working classes could command four times as much of the necessaries of life then as they can now; and this fact is borne out by the evidence of Sir John Fortesque. "The people," says he, "have plenty of fish, flesh and fowl, the best furniture in their houses, they are well clad in woolen clothes; they never drink water except in Lent, or fast days, but wine or beer;" yet these are the times that are called dark and barbarous. It would be well, indeed, if the barbarous custom of having enough of the comforts of life might again be the lot of the laboring population of England.

Contrast the above statement with the speech of the Queen of England from the throne, in the early part of 1846. "I deeply lament the failure of the potato crop, as this is an article of food that forms the chief subsistence of great numbers of my people."

And again, on the 19th of January, 1847, her speech from the throne commences thus.

"My Lords and Gentlemen—It is with the deepest concern that upon your again assembling, I have to call your attention to the *dearth of provisions*, which prevails in Ireland, and in some parts of Scotland."

How the Queen could make such a declaration in the face of the civilized world, when it is a well known fact that the same people who are living upon potatoes—nay even dying by thousands for want of the necessaries of life, are exporting annually several thousand tons of pork, grain, poultry, eggs, butter, cheese, and many other articles of food which their insufficient remuneration for labor will not allow them to touch—I say, how she could come before Parliament and make this statement, I cannot imagine.

Innumerable facts might be quoted in favor of the principles which I have endeavored here to inculcate; but I leave what has been said to the reflection of the candid reader.*

---

* A recent number of the London "Times" Newspaper, contains the following paragraph.

"Poor Ireland exports more food than any other country in the whole world—not merely more in proportion to its people, or its area, but absolutely more. Its exports of food are greater than those of the United States, or of Russia, vast and inexhaustible as we are apt to think the resources of those countries are. Such a fact as this is very compatible with a people being poor; but it at least shows that one ought to inquire what sort of a poverty it is. Stand on the quays of Ireland, and see the full freighted vessels leaving her noble rivers and coves. You will there see, that, so far from Ireland being utterly, radically and incurably poor, barren and unprofitable, she is one of the great feeders of England; nay, its chief purveyor. Ireland does this out of her poverty, besides feeding, after a manner, an immense population. It is this that adds so painful an interest to her miserable state; that she should 'make many rich,' and yet remain herself so poor, and be the author of an abundance which she is

LETTER IV.

AGRICULTURAL LABORERS OF WILTS, DORSET, DEVON, AND SOMERSET.

I have been very careful to collect the facts here set down from the most correct sources; and have inserted nothing but what was confirmed by my own experience and investigation. In this and future letters I shall treat each branch of industry separately. And first of agriculture.

The agricultural districts of England are widely separated from each other; the nature of the employment, the remuneration for labor, habits of the people, &c., are also different in the several districts, and it will therefore be necessary for a clear understanding of the subject, that I should take each district separately. I shall confine myself in this letter to the four counties of Wilts, Dorset, Devon and Somerset.

Strictly speaking, there is no great uniformity in the agricultural features of these contiguous counties, al-

---

not permitted herself to enjoy. Great Britian not only draws nothing from the Irish treasury, but gives Ireland the gratuitous benefit of her own enormous revenue. The whole of the Irish revenue, including every sixpence obtained for customs, excise, stamps and postage—from tea, sugar, coffee, tobacco, spirits, and from every other article imported, as manufactured in the island, is spent in Ireland itself. Not one sixpence is remitted to the British exchequer. In point of fact, the current had already set in from the British to the Irish treasury. Ireland, then, is, at the same time, rich and poor. It produces a vast superabundance of food, but that food is drained from its shores. It is not, however, drained by the state. It is drained in a great measure, by the landlords and their creditors, who, the more they can get, the more they will drain."

This at least shows, that the famine in Ireland is not the result of the Providence of God; but the mismanagement of the rulers of the land.

though there is sufficient in their practice, customs and peculiarities, to justify the classification here adopted. They are all more or less dairy and grazing counties; that is, devoted to the manufacture of cheese, rearing of young cattle, and sheep farming. Pigs are also reared in vast numbers, and constitute an essential appendage to the dairy farm; and in the low tracts, immense flocks of geese and other fowls are annually fatted for the city markets. The most extensive orchards in Britain are to be found in the valleys of Devon and Somerset, and hence the management of these, and the subsequent manufacture of cider and perry, constitute one of the main duties of the farmer.

The practice of employing women prevails more or less in all these counties; their out-door labor consists in hay-making, reaping, hoeing turnips, weeding corn, picking stones, beating manure, planting and digging potatoes, pulling turnips, and occasionally hacking them for cattle. They are also sometimes employed in winnowing corn, about the threshing machine, and in leading horses and oxen at the plough. The in-door labor is milking and making cheese, and looking after, cleaning, turning, weighing, and removing the cheeses that are already made. This is a sort of work that is said to be "never finished."

The wages of women differ slightly, not only in adjoining counties, but even on different farms, according to the character of the farmer, or the ability and skill of the laborer. Generally speaking, all light work, such as apple-picking, turnip-hoeing, stone-gathering, and the like, is paid at the rate of sixteen cents a day, with an allowance of *cider*. Hay-making at twenty cents, and potato lifting and harvest work at twenty-four cents, with *dinner in harvest time*, and a quart of cider. Most of the cider is saved by the women for their husbands. When women work by the piece, they strive to earn higher wages, say $1,25 a week in summer, $1,00 to $1,10 in winter. These are the regular wages *with cider*.

In some of the Dorsetshire villages, the younger females are much engaged in button-sewing, and as it is a lighter employment, are not tempted to field work, unless during summer, and then only at twenty to twenty-four cents a day.

Women accustomed to field labor, represent it as good for their health and spirits; this, however, must be taken with some restrictions; for where women poorly clad are exposed to cold and wet, and this for ten or twelve hours a day when the weather will permit, catarrhs and rheumatism will be the result.

From this cause we find them complaining, as their husbands too often do, of stiffness and pains at the joints, long before such complaints can be the result of old age, or natural infirmity.

Regarding the moral condition of the females in these counties, the evidence is very conflicting. Here we find a clergyman inveighing against field labor, as the source of most of the immorality in the district; another, an old fashioned farmer, declaring quite the contrary; a third, less biased than either, admitting that field labor is not the best school for morals; a fourth, a grave old man, says, "those young ones would never stick to their work were it not for the cider I find them, and the fun they make for themselves."

There are three modes in which the employment of children may take place within these counties: they may be taken to assist their parents, may be hired by the day or week as women are, or may be apprenticed by the parish. The servitude, in the case of agricultural apprenticeships, extends from the age of ten to twenty-one for boys, and generally till marriage for females. The younger girls are employed in the farm houses to look after children, and to do other light work. Boys from seven years of age (I have seen them even younger,) to twelve are employed in bird-scaring, taking care of poultry, following the pigs in the acorn season, herding cattle, getting wood for the house, and the like. As they get stronger they lead the horses and oxen at the plough, make hay, and hoe turnips, and by fourteen or fifteen years they begin to hold the plough, attend to the stable, help the carter and drive the team. After that time they commence mowing, reaping, hedging, ditching, and the other difficult operations performed by the farm laborers. The hours of labor for boys are the same as for men and women; their wages are from thirty-six cents to $1,10 a week, with a pint of cider a day. They are taught to love drink from their earliest age, and a few years so confirms them in the taste, that they rarely, if ever, get rid of it in after life.

Such is an outline of the employment of women and children in agriculture in these counties, and the effect which it is calculated to produce upon their physical and moral condition. The labor, taken by itself, would seem to be comparatively harmless; but, taken in connection with the general condition of the laborer, it tends in a great degree to depress that which is already by no means exalted. The early age at which most of the children are taken from school, prevents their getting even the rudiments of education. They are too early associated in promiscuous labor with men and women, in whose vices they become adepts long before they have attained the years of maturity.

The almost constant employment of women in the fields has many bad effects upon their families; their cottages are not properly attended to, their children are neglected, clothing is allowed to get dirty and torn, and many matters in domestic economy are allowed to fall into disorder, so much so, that some women say it is more to their advantage to stay in and attend to affairs at home. It is, however, to the general condition of the agricultural laborer in these counties that we are to look for the main evils that are said to affect his case. His wages vary from two to three dollars per week. This is inadequate when he has a large family to support; and the consequence is, want of sufficient clothing, neglect of personal cleanliness, and scantiness of diet. At the farm houses, where the single men live with their masters, of course the fare is better; it is by the married cottagers that the greatest evils are felt.

Their cottages are small and in bad order; they are generally damp and in a state of decay; there is no inducement to cleanliness or neatness on the part of the laborer, and hence what ought to be *homes*, are mere hovels for shelter. Cottages generally have two apartments; a great many have only one. The consequence is, that it is very often extremely difficult, if not impossible, to divide a family so that grown up persons of different sexes do not sleep in the same room. Three or four persons not unfrequently sleep in the same bed, and in a few instances I have heard of families who have arranged it so, that the females of both families slept altogether in one cottage, and the males in the other. Generally an old shawl is suspended as a curtain between two beds in one room.

LETTER V.

AGRICULTURAL LABORERS OF KENT, SURREY AND SUSSEX.

The reader's attention is called in this letter to the condition of the agricultural laborers of the counties of Kent, Surrey and Sussex. I may here remark that I have resided in various parts of these counties, and superintended one estate of about twenty acres of land, and that many of the facts here related have come under my own observation.

The agriculture of these counties differs in many respects from that of other districts in England. These counties present a great variety of external features, when taken separately; but when collected and compared together, they exhibit a remarkable unity and sameness. The great formation of the *wealden clay*, the *sand* and the *chalk*, belong to each and all. This large and central tract of country is girt with a belt of chalk hills, a fringe of sand forms the union between the chalk and the wealden.

The employment of women is not so varied and promiscuous in these as in other counties of England. Generally speaking, there are few grain-growing or stock-rearing districts; hence corn, hay, turnip and potato work, is by no means common. Occasionally we find them at the hay-harvest, picking stones from the meadow land, dropping beans, or hoeing turnips, but very rarely at reaping, or potato lifting. The winter work is performed by the men and boys kept on the farms. The chief employment of the women is in the *hop gardens* and *orchards*, and in the former there is continuous work for the greater part of the year.

As the culture of the hop is peculiar to this district, it may be necessary to give a brief description of the mode in which it is conducted. The land is prepared with considerable attention by fallowing, deep stirring, and cleaning; it is next thrown into rows of little hillocks at equal distances; and in these hillocks the young shoots of the hop (previously nursed in the orchard) are planted.

Opening the hills consists of digging a hole about two feet square and two feet deep in the centre of these hillocks; this is done for the purpose of loosening the soil, and depositing the manure for the future crop. When the soil has been replaced, and the hillocks again completed, several young shoots are planted. This is performed in February and March by men and boys. ***Poleing*** is the next process, which is performed in April, before the hop begins to shoot. The pole, or stake, from eight to ten feet high, furnishes a support to the climbing vine of the plant; several are fixed in one hill, and it requires strong muscular exertions to do so. It is invariably performed by the man, who, however, receives assistance from his wife, his son, and sometimes his daughter. ***Tying*** is the next process; that is, fastening the climbing vine to the poles. This process is carried on from the moment the vine has shot above ground, to the time that it arrives at maturity.

Tying is invariably done by the women, who are occasionally assisted by children. When the hops reach the top of the poles, the women have to mount on a kind of *ladder*, which enables them to fasten the vines which may have blown off; this is called *horseing*. *Skimming*, is effected by an instrument, so called, drawn by horses between the rows of hillocks for the purpose of loosening the earth and weeds. The horses are carefully led by boys, and the instrument is guided by men.

When the seed of the hop is ripe, which generally

takes place in September, the *picking season* commences; and in this process women and children of all ages are employed. It is necessary that it be carefully and speedily done, hence the great annual influx of Irish and Londoners to the hop districts. Thus it will be seen that hop growing is one of the most difficult and laborious duties of the farmer, its culture requiring an all but ceaseless round of watchfulness and toil.

In some parts of Kent, there are many thousand acres of *orchard* land, and in these, women and children are much employed in weeding, gathering fruit, and the like.

We very rarely hear of *cider* being allowed by the farmer; the practice is almost unknown in these counties.

The hours during which female labor is continued are variable in these counties, owing to the almost universal practice of doing all sorts of work by contract; that is, at so much per acre, per bushel, and so on. The time of work and meals are fixed by the laborer, who is naturally anxious to earn as much as possible. We may, however, mention from ten to twelve hours per day as the most general.

With regard to wages, it is still more difficult to strike an average, though we may mention sixteen, twenty, and twenty-four cents a day, for females, on arable farms; twenty to twenty-four cents in orchards; twenty-four to thirty-six on harvest fields; and from twenty-four to forty-eight cents in hop plantations, according to the skill and ability of the worker. The men and boys earn much the same as mentioned in my last communication. The employment of children, especially boys, is more common in this, than in any other district, owing to the abundance of light work which can at all times be easily obtained. They generally commence at seven or eight years of age, and continue at this work till twelve.

The effect of juvenile labor upon health is not much

complained of, although it may be observed that a weakness of limb, great turning out of the feet, and a draggling gait, is common to most of the boys, owing to being set to labor at too early an age.

Rheumatism is the chief disease complained of by men and women, arising from exposure to wet and cold, a want of cleanliness, and inadequate clothing and diet.

On the arable and woodland districts there is nothing peculiar in respect of morality; but in the hop and orchard localities, the morals of the work people is far from being well spoken of; and the cause generally assigned is this.

At the proper season, *hop-pickers* come from all parts of England and Ireland, and amongst them may be found unfortunate members of various classes. Great numbers go from the crowded districts of London, and they are the most vicious and refractory. These associate promiscuously together during the day, and are for the most part, *herded* together, if I may use the term, during the night, so long as the season of hop picking continues.

Ignorance prevails to an alarming extent among the resident laborers in this district. The school masters say that two-fifths of their scholars are regularly absent. It is quite common to meet with boys engaged on farms who cannot read or write. I have had boys in this state of ignorance working for me, and it is remarkable how eagerly they avail themselves of any favorable opportunity of learning, *when proper encouragement* is held out to them. The being of a God, a future state, the number of months in the year are not universally known. Superstition, the result of ignorance, in this case at least, still lurks among the laboring classes in these counties. The belief in charms for healing of bodily hurts is not uncommon.

The agricultural laborer in this district holds at present

a low position in the social scale; and the urgency of immediate wants, and the desire to keep out of the workhouse, compel him too frequently to drive his children to labor at an early age. It is a distressing fact, that the wife and children are *obliged* to accompany the husband in his labor in the fields. The calling of the mother away from the charge of her household, and the intrusting domestic matters to very young girls, are attended with consequences which the reader may imagine better than I can describe. The cottager cannot be said to have a comfortable home. The common practice of keeping as many lodgers as can be crammed into an apartment during the hot season, does not improve the health or comfort of the inmates. While the food and clothing of the laboring classes in this district is much better than in some others, it is by no means such as these industrious people ought to be able to obtain in return for their industry. But while the laborer has much to answer for his own improvidence, his master can by no means be exempt from a share of the evil. In addition to a want of comfortable cottages, which every landlord ought to feel it his duty and interest to provide, the laborer is half robbed of his scanty earnings by the *truck* system.

It is a common practice in these counties to pay the wages, both of men and women, by a check drawn upon the miller of the village, who is generally related to the farmer, the laborer getting part of his wages in flour and part in money; or, it may be, that the miller again hands him over to the grocer; and thus the poor man in general pays from 25 to 30 per cent more for his victuals than in justice and honesty he should. Nor is this all; he pays the highest price for the worst goods, and dares not complain. This crying evil ought to be removed.

LETTER VI.

IGNORANCE AND SUPERSTITION IN KENT.

In order to give the reader a clearer view of the state of the people in this district, I will here relate some remarkable transactions which took place in May, 1838, near Canterbury, in Kent. I was then living about thirty miles from the scene of action, and well remember the sensation these events produced in the public mind.

Kent is one of the most beautiful counties in England, and the villages and scenery around Canterbury are peculiarly English. Gently rising hills and picturesque vales, covered with a rich herbage, all giving proof of a minute and skilful husbandry, succeed to each other. Fields of waving corn are interspersed with gardens, hop grounds and orchards.

The hero of the Kent disturbances, was John Nicolls Thoms, the son of a small farmer and maltster, at St. Columb, in Cornwall. He appears to have entered life as cellarman to a wine merchant in Truro. Succeeding to his master's business, he conducted it for three or four years, when his warehouse was destroyed by fire, and he received about $15,000 in compensation from an insurance company. Since then, during more than ten years, he had been in no settled occupation. In the year 1833, he appeared as a candidate, successively for the representation of Canterbury and East Kent. His fine person and manners, and the eloquent appeals he made to popular feeling, secured him a certain degree of favor; but were not sufficient to gain his object. Though baffled in this, he continued to address the populace as their peculiar friend, and kept up his influence among them. In July of the same year he made an appearance in a court of law on behalf of the crew of a smuggling vessel, when he conducted himself in such a way as to incur a charge of perjury.

He was consequently condemned to transportation for seven years, but, on a showing of his insanity, was committed to permanent confinement in a lunatic asylum, from which he was discharged a few months before his death, on a supposition that he was of sound mind.

Immediately after his liberation, he resumed his intercourse with the populace, whose opinion of him was probably rather elevated than depressed by his having suffered from his friendship for the smugglers. He repeated his old stories of being a man of high birth, and entitled to some of the finest estates in Kent. He sided with them in their dislike of the new regulations for the poor, and led them to expect that whatever he should recover of his birthright, should be as much for their interest as his own. There were two or three persons of substance who were so far deluded by him as to lend him considerable sums of money. One gentleman loaned him $1,000 on some supposed title deeds.

Latterly, pretensions of a more mysterious nature mingled in the ravings of this madman; and he induced a general belief amongst the ignorant peasantry around Canterbury, that he was either the Saviour of mankind sent anew upon earth, or a being of the same order and commissioned for similar purposes. He took the title of SIR WILLIAM PERCY HONEYWOOD COURTENAY, Knight of Malta, and King of Jerusalem. One of his deluded followers declared afterwards that he could turn any one that once listened to him whatever way he liked, and make them believe what he pleased. He was very kind to the poor, and would give the last shilling in his pocket to a poor man. His aspect was very imposing. His height about six feet, his features were regular and beautiful, a broad, fair forehead, aquiline nose, small mouth, and full, round chin. The only defect was a somewhat short neck. He possessed uncommon personal strength.

Some curious significations of the enthusiam he had excited were afterwards observed in the shape of scribblings on the walls of a barn, which I copy verbatim. "If you new he was on earth, your harts Wod turn." "But don't Wate too late." On the side of a barn door was the following :—"O that great day of judgment is close at hand." "It now peps in the door every man according to his works;" "our rites and liberties We Will have."

On Monday, the 28th of May, the frenzy of Thoms and his followers seems to have reached its height. With twenty to thirty persons, in a kind of military order, he went about for three days among the farm houses and villages in the vicinity of Canterbury, receiving and paying for refreshments. One woman sent her son to him, with a "mother's blessing," as to join in some great and laudable work. He proclaimed a great meeting for the ensuing Sunday, which he said was to be "a glorious but bloody day."

At one of the places where he ordered provisions for his followers, it was in these words, "feed my sheep." On another occasion he went away from his followers with a man of the name of Wills, and two other of the rioters, saying to them, "Do you stay here, whilst I go yonder," pointing to a bean stack, "and strike the blow."

When they arrived at the stack, to which they marched with a flag, the flag bearer laid his flag on the ground, and knelt down to pray. The other then put in a lighted match, which Thoms seized and forbade it to burn. This on their return to the company was announced as a miracle.

On Wednesday evening they stopped at the farm house of Bossenden, where the farmer finding that his men were seduced by the impostor from their duty, sent for constables to have them apprehended. Two brothers, named Mears, and another man, accordingly went next morning, but on their approach Thoms shot one of the brothers dead with a pistol, and aimed a blow at the other with a dagger, whereupon the two survivers fled.

At an early hour he was abroad with his followers, to the number of about forty. He undertook to administer the sacrament, in bread and water, to the deluded men who followed him. He told them, on this occasion, as he did on many others, that there was great oppression in the land, and throughout the world; but that if they would follow him, he would lead them on to glory. He told them he had come to earth on a cloud, and that on a cloud he should some day be removed from them; that neither bullets nor weapons could injure him or them, if they had but faith in him as their Saviour; and that if ten thousand soldiers came against them, they would either turn to their side, or fall dead at his command. At the end of this harangue, Alexander Foad, whose *jaw* was afterwards shot off by the military, knelt down at his feet and worshipped him; so did another man of the name of Brankford. Foad then asked Thoms whether he should follow him in the body, or go home and follow him in heart; to which he replied, "Follow me in the body." Foad then sprang on his feet, in an ecstasy of joy, and with a voice of great exultation, exclaimed, "O, be joyful! O, be joyful! The Saviour has accepted me. Go on, go on; till I drop, I'll follow thee!" Brankford also was accepted as a follower, and exhibited the same enthusiastic fervor. At this time his denunciations against those who should desert him, were terriffic. His eyes gleamed like a coal of fire while he was scattering about these awful menaces. It is believed that if any of his followers had attempted to desert him at this time, he would have shot them. A wood-cutter, (not a follower,) went up to him, shook hands, and began to converse with him, and among other things, asked him if it was true that he had shot the constable. "Yes," said he coolly, "I did shoot the vagabond, and I have eaten a hearty breakfast since. I was only executing upon him the justice of heaven, in virtue of the power which God has given me."

The two repulsed constables had immediately proceeded to Feversham, for the purpose of procuring fresh warrants and the necessary assistance. A considerable party of magistrates and other individuals, now advanced to the scene of the murder, and about mid-day (Thursday, May 31st, 1838,) approached Thoms' party, at a place called the Osier-bed, where the Rev. Mr. Handly, the clergyman of the parish and a magistrate, used every exertion to induce the deluded men to surrender themselves, but in vain. Thoms defied the assailants, and fired at Mr Handly, who then deemed it necessary to obtain military aid before attempting further proceedings. A detachment of the 45th regiment, consisting of 100 men, was brought from Canterbury, under the command of Major Armstrong. A young officer, Lieutenant Bennett, who belonged to another regiment, and was at Canterbury on furlough, proposed, under a sense of duty, to accompany the party, on the condition that he should be allowed to return before 6 o'clock to dine with some friends.

At the approach of the military, Thoms and his men took up a position in Bossenden Wood, between two roads. Major Armstrong divided his men into two bodies, of equal numbers, that the wood might be penetrated from both of these roads at once, so as to inclose the rioters; the one party he took command of himself, and the other was placed under the command of Lieutenant Bennett. The magistrates, who accompanied the party, gave orders to the officers to take Thoms dead or alive, and as many of his men as possible. The two parties then advanced into the wood by opposite roads, and soon came within sight of each other, close to the place where the fanatics were posted. A magistrate in Armstrong's party endeavored to address the rioters, and induce them to surrender; but while he was speaking, Lieutenant Bennett had rushed upon his fate. He had advanced, attended by a single private, probably for the purpose of calling upon the insurgents to submit, when the madman who led them advanced to meet him, and Major Armstrong had just time to exclaim, "Bennett, fall back," when Thoms fired a pistol at him within a few yards of his body. Bennett had apprehended his danger, and had his sword raised to defend himself from the approaching maniac; a momentary collision did take place between him and his slayer, but the shot had lodged with fatal effect in his side, and he fell from his horse a dead man. Thoms fought for a few seconds with others of the assailants, but was prostrated by the soldier attending Mr. Bennett, who sent a ball through his brain. The military party then poured in a general discharge of fire-arms on the followers of the impostor, of whom eight were killed, and others severely wounded, one of whom afterwards died. A charge was made upon the remainder by the surviving officer, and they were speedily overpowered and taken into custody.

Of the deluded men who followed Thoms, nine were killed, who left four widows and ten children; sixteen were sent to jail, and eleven discharged on bail. Nearly the whole were men of steady, reputable character, and some of them were in the receipt of wages considerably above the average of the district.

This occurrence broke upon the public ear with a startling effect, and the Central Society of Education in London sent a gentleman down to investigate the circumstances on the spot. The result of his inquiries has been given to the world in an elaborate paper in the third volume of the publication issued by the Society, to which book I refer the reader for a further account of these riots. I will, however, make a few short extracts to show the state of education among the peasantry of Kent.

The report gives to fifty-one families examined, forty-five children above the age of fourteen years, and 117 under that age. Of the first class, eleven only can read and write, twenty-one can read a little, and the remainder cannot. In the second class, forty-two attend school, but several of these go only occasionally, the rest do not go at all. Six only can read and write; of twenty-two who can read, only thirteen read fluently, and nine very little; and the remainder cannot read at all. In twelve families the *boys* assist their father in his labor, and seldom receive any instruction. In fifteen families the *girls* do the household matters, and in thirty-four families they do nothing but wash and needle-work.

The parish possessed a Sunday school, and three others, in one only of which was writing taught. This school was kept by a master, who, being from physical infirmity incapable of labor, was obliged to adopt this mode of life. He had only eighteen scholars, and half of this number came from neighboring parishes. The two other schools were merely dame-schools, in which nothing but sewing and reading are taught. Many of the children attend so irregularly, and are often absent for such long periods, that they forget all they have learned. Owing to this, some children are unable to read, after being members of the school two or three years. The gentleman above mentioned, says, "It would be easy, if it were required, to adduce reasons for believing that the gross ignorance shown to exist in these districts, is not confined to them, but that their condition may be regarded as a fair sample of that of the same class in other parts of the country." And again, "A little consideration of the nature of rural life will show the danger of leaving the peasantry in such a state of ignorance. In the solitude of the country, the uncultivated mind is much more open to the impressions of fanaticism than in the bustle and collision of towns. In such a stagnant state of existence the mind acquires no activity, and is unaccustomed to make those investigations and comparisons necessary to detect imposture. The slightest semblance of evidence is often sufficient with them to support a deceit which elsewhere would not have the smallest chance of escaping detection. If we look for a moment at the absurdities and inconsistencies practised by Thoms, it appears at first utterly inconcievable that any person out of a lunatic asylum could have been deceived by him. That an imposture so gross and so slenderly supported should have succeeded, must teach us, if any thing will, the folly and danger of leaving the agricultural population in the debasing ignorance which now exists among them."

Such is a brief outline of one of the most strange and singular popular delusions of modern times. It would have given me pleasure to have been able to say, that a great improvement had taken place since 1838; such, however, is not the case.

LETTER VII.

AGRICULTURAL LABORERS OF SUFFOLK, NORFOLK, LINCOLN, YORKSHIRE AND NORTHUMBERLAND.

Leaving the counties of Kent, &c., let us now proceed northward, into the counties of Suffolk, Norfolk and Lincoln. These counties are distinguished from other districts of England, in consequence of their being more exclusively under tillage. The business of mixed husbandry—that is, the production of almost every variety of *white* and *green* crop—is here carried to great perfection; this involves a constant routine of manual labor, which the custom of the country consigns chiefly to women and children. As in the counties already noticed, we find the people engaged in field labor occasionally suffering from bad colds and severe rheumatism. The employment of married women is much lamented, as it takes them away from their domestic duties, and leaves their cottages and children in a neglected state. But the crowning evil in these counties is "the gang system."

This is a method of working which had its origin at Castle Acre, in Norfolk, about twenty-five years ago, and now prevails in many contiguous parishes. Mr. Denison, a Government Commissioner, thus describes the system:

"Suppose a farmer wishes to have a particular piece of work done, which will demand a number of hands, he applies to a gang-master, who contracts to do the work, and furnish the laborers. He accordingly gets together as many hands as he thinks sufficient, and sends them *in a gang to their place of work.* If the work, as usually happens, be such that it can be done by women and children, as well as men, the *gang* is in that case composed of persons of both sexes, and of all ages. They work together, but are superintended by an *overseer*, whose business it is to see that they are steady to their work, and to check any bad language or conduct. The overseer usually goes with the gang to the place of work, and returns home with them when they leave off for the day."

The system is said to be productive of the worst consequences, which will be readily admitted, when it is considered that the gangs are generally composed of the lower orders from the towns, yoked together without regard to age, sex or character, and crowded together at night when the distance compels them to lodge on the farm. There is a complete disseverment between the farmer and the laborer; the former has no interest either in the character or condition of the latter; the whole power, as well as responsibility, is delegated to an ignorant and grasping gangsman, whose tyranny is the more oppressive, that he is little if at all superior either in intellect or station to the laborer.

Such a practice as this has no necessity to justify, no single advantage to recommend its continuance. The gang-master may find it a profitable affair, and to the farmer of 600 acres it may prove an easier mode of getting his work performed; but it seems to be at once injurious to the interest and morals of the laborer.

It subjects him in the first place, to the truck-trading oppression of the gang-master, who not only screws down his wages to the lowest cent, but supplies him with inferior articles at the highest price; while it subjects him to greater personal toil, and contaminates the morals of his children. There is nothing in the agricultural peculiarities of this district which demands the maintenance of such a system.

With respect to education, it is not much better than in the counties we have just left. In the spring of the year 1842, I was engaged by a wealthy Baronet to take the management of a school in this district, and I well remember he stated to me, in conversation upon this subject, "that the children in his parish were as ignorant as brutes." An unforeseen occurrence prevented me from fulfilling that engagement.

Let us now pass on to Yorkshire, Northumberland, &c. In this district the women and children work in the fields, under much the same circumstances as in most other rural situations, and for much the same remuneration. There is in the East Riding of Yorkshire, a mode of paying wages which is considered a great evil, and may be thus described. The male laborers are fed in the farm-houses, and have a certain proportion of wages deducted to pay for their meat. This sum, (twenty-four cents a day,) amounts to nearly one half his whole wages; so that setting aside her husband's food, about $1,50 to $1,75 is all that a woman has with which to confront the rest of life; her food, that of her children, the rent of the cottage, fuel, schooling, clothing, medical attendance, and taxes, have all to be provided for out of this sum.

The farmers like this system, either because they profit by it, or because they have a notion (which is very reasonable,) that men work better with a *full* belly than an *empty* one. The men like it, because, no doubt, they get a better dinner than would otherwise fall to their share; but upon the women and children it must operate as an evil

In the more prosperous districts of Yorkshire, many of the cottages have small gardens in which most of the vegetables for the family can be reared, and some are permitted to have a cow at grass. These, however, are exceptions. In point of mere victualling and personal comfort, the Yorkshire peasantry may be said to be comparatively well off; but there still seems ample room for the improvement of their condition in education and more comfortable housing.

Farm servants in Northumberland, are engaged upon a system different from that which prevails in other parts of England. In the absence of villages (which are rare) to supply occasional assistance, each farm must depend upon its own resources; a necessity is thus created for having a disposable force of boys and women always at command, which is effected in the following manner. Each farm is provided with an adequate number of cottages, having small gardens adjoining, and every man who is engaged by the *year* has one of these cottages; his family commonly find employment, more or less, but *one female laborer he is bound to have always in readiness* to answer the master's call for assistance, and to work at stipulated wages. To this engagement the name of *bondage* is given, and such female laborers are called *bondagers;* or women who work the *bondage.*

Of course, where the *hind* (as such yearly laborer is called) has no daughter or sister competent to fulfil for him this part of his engagement, he has to take his wife to do it, or hire a woman servant; and this, in some sense of the word, may be a hardship to him; but, in the first place, this is not very common; and in the second, the advantages of the system, even with this drawback, are unquestionable. The wages of the laborer are paid chiefly in the produce of the farm, viz: in addition to a cottage and small garden, he has a certain quantity (as per agreement) of wheat, rye, barley, oats, peas, beans, potatoes, wool, &c., with about $20 in money during the year, and the keep of a cow. The wages of the women and children are paid chiefly in money. This system seems to be the best of any in England for agricultural laborers, and is deserving of all the commendation which the practical farmers of Northumberland unite in bestowing upon it.

It is pleasant to hear that the education in Northumberland is good, that the people eagerly seek to acquire knowledge, and that it is a rare thing to find a grown-up laborer who cannot read and write, and who is not capable of keeping his own accounts. Such a state of things contrasts favorably with the neglected condition of more southern counties; and when we are told that this education is not obtained through national schools, charitable institutions, and the like, but by the exertions of the peasant himself, it indeed bespeaks a state of society where sobriety is habitual, and intelligence held in high estimation.

There is not much said against the morality of this district, but it must be borne in mind that this is comparatively a thinly inhabited county, and that there is no inducement for people to come here in search of employment from other parts of the country; consequently the farms are not crowded with the superfluous population of large towns, like the hop-grounds and orchards of Kent.

The question may here be asked, are those relations which ought to subsist between the employer and employed, honestly attended to? Does property discharge its *duties* to those by whose arm it is rendered productive? To this question I regret to find that all the evidence before the public, replies in the negative. Each district exhibits in a greater or less degree the neglected state of the peasant; there is little or no provision made for the proper education of his children, and equally scanty attention is paid to the fostering of sober and industrious habits. Instead of the smiling cottage garden, which we naturally associate with our ideas of "Merrie England," we are told of, and see in many places, miserable hovels, with accommodation so niggardly allotted, that parents, sons and daughters, must eat, sit and sleep in one apartment! From Kent to Northumberland, these evils are complained of; and if hovels, which would be thought unfit housings for dogs and horses, are to be the reward of "industry embrowned with toil," we may cease to wonder if that industry should degenerate into careless and improvident habits, and that the workhouse, with its "regulation diets" and "ventilated halls," should be preferred to the cottage hearth, and the home of an honorable self-dependence.*

The London Daily News of November 30th, 1846, contains an account of the Taunton Agricultural Society, from which I make the following extracts.

Sir Alexander Hood presided on this occasion, and among his supporters was Sir Thomas Lethbridge; whose speech on proposing the health of the chairman, was the leading feature of the festival. The latter part of this truly admirable speech I give at length, as it will show the reader that the subject of insufficient remuneration for labor, is beginning to occupy the attention of all classes.

The venerable baronet spoke at great length of the advantages of agriculture, and then drew the attention of his hearers to the farm laborer, as follows.

"There is one thing I wish to tell you. You have gone far enough in fattening stock. It strikes me that you are putting too much fat on lean bones. That does no good to anybody. We have shown what we can do, and you can do what you have done, over and over again. But it is not wise and profitable to do it. If it takes away an atom of food from the poor, stop it. Breed as much as

* Those persons who may wish for further information upon the subject of agricultural labor, may consult the reports of the Government Commissioners.

you like; breed to the top of the tree. Breeding in all branches! (Cheers.) Go on with your breeding, gentlemen, but don't put too much fat upon lean bones. You are starving people by doing so. One of my tenants told me last week that he had two beautiful oxen for which he has got a prize. They were brought here to-day. They were fed with ten times as much as they ought to have been. Is this wise? Where is your money? It is not in your pocket. It is on the animal's back. One half of it will go into the chandler's shop, and into the soap-basin. Who will thank you? You have shown what you can do. See if you cannot do something better. I meant to say a great deal on the subject of the laborers, but I have not breath enough to do it well. But I must say this, that you and I may take counsel together and spend money; but if you forget the men whose labor gets all for you, you are acting ungratefully. I therefore, in three words say, '*Raise your wages.*' I have done it myself, and I tell all my friends to do it. I tell you all to raise your wages, and I tell you this—there never was a man in your employ who ever struck a stroke for his master, but struck that stroke with redoubled force and energy (and if he does this, who profits?) when he knows he is well used. (Cheers.) How is he used? In Somerset they give what? (A voice, 'They give what it is worth.') I guessed I should be interrupted, but recollect my grey hairs. Recollect my experience. Recollect my triumphs in farming. Recollect how many laborers I have employed—how I have lived among them. Recollect, also, that though lower in the scale than some are, they are men before God as well as yourselves. (Cheers.) Woe be to him who will not mete out to others what he wishes to have meted out to himself. But must you do it if the rules of political economy say you should not do it? I say, 'Avaunt, political economy,' if they say so. But, gentlemen, the rules of political economy say no such thing. They do not make men brutes. They do not seal up and freeze all the finer feeling of human nature of man toward man. They have thrown open the whole world to competition, and I hope and believe that out of this the best results will come to all. I believe that those principles are true. (Cheers.) So far as we have tried them, they appear to be true. No man can predict what they may do for us in the future. But looking, as I have always looked, and always do look, I am well convinced, as surely as I am standing in this room—most likely for the last time—(Loud cries of 'No, no,' and vehement cheering) I firmly believe we are going on in the right path—to the path which leads to the prosperity of the landlord, the prosperity of the laborer, and to the prosperity of the whole people, and even more than all, to permanent peace; and let me say, God grant it may be so, for peace is the greatest blessing that can flow to a people like the English, who are united among themselves, and are beloved by the whole world because they have made the rule of right, their rule. (Cheers.) Now, sir, I have one more word to say, and then I will sit down. It is this : I shall offer a higher reward if the society will not, but the society had better do it, not for putting fat upon lean bones, but for putting *comfort into the cottages of the poor laborers.* (Cheers.) If you do not encourage them, all your meetings here are good for nothing. You may as well go home when you leave this, and say, 'Well, I have done very little.' (Laughter.) What I propose, is, to give to the farmer who from the first day of December, 1846, to the first day of November, 1847, shall employ the greatest number of laborers and servants, at the highest rate of wages, by the week or month, and without reference to the size of the farm, the sum of £15. ($75.) (Cheers.) I hope the society will do more than it has yet done for the laborer. He *must* be raised from his present poverty and degradation, and I call upon you to do what you can towards accomplishing that purpose."

LETTER VIII.

THE ENGLISH FACTORY SYSTEM—ITS EARLY HISTORY.

I wish it to be distinctly understood, that any thing I may advance in this or other letters, has no reference to the factory system of *New* England. Upwards of thirty years' experience and observation teaches me that the factory system of *Old* England and that of *New* England are as widely different in their results, as the two forms of government under which they are respectively carried on. Not to enter into details at present, I would simply say that the system of Old England could not exist in New England; neither could the New England system exist in Old England; the thing is impossible, so long as the systems of government remain as they are. With this preliminary observation I will proceed to make some extracts from printed documents, as the most satisfactory method of showing the actual character and consequences of the English Factory System.

"It may not be amiss to inquire how it came to pass originally, that, in England, always boasting of her humanity, laws were necessary in order to protect little children from the cruelties of the manufacturer.

It is well known that Arkwright's (so called, at least) inventions took the manufactures out of the cottages and farm-houses of England, where they had been carried on by mothers, or by daughters under the mother's eye, and assembled them in the counties of Derbyshire, Nottinghamshire, and, more particularly, in Lancashire, where the newly invented machinery was used in large factories built on the sides of streams, capable of turning the water-wheel. Thousands of hands were suddenly required in these places, remote from towns; and Lancashire, in particular, being till then but comparatively thinly populated and barren, a population was all she now wanted. The small and nimble fingers of little children being by very far the most in request, the custom instantly sprang up of procuring *apprentices* from the different parish workhouses of London, Birmingham, and elsewhere. Many, many thousands of these little hapless creatures were sent down into the North, being from the age of seven, to thirteen or fourteen years. The custom was for the master to clothe his apprentices, and to feed and lodge them in an "apprentice-house" near the factory; overseers were appointed to see to the works, whose interest it was to work the children to the utmost, *because their pay was in proportion to the quantity of work done.*

"Cruelty was, of course, the consequence; and there is abundant evidence on record, and preserved in the recollections of some who still live, to show that in many of the manufacturing districts, but particularly, I am afraid, in the guilty county to which I belong, cruelties the most heart-rending were practised upon the unoffending and friendless creatures who were thus consigned to the charge of master-manufacturers; that they were harassed to the brink of death by excess of labor; that they were flogged, fettered and tortured, in the most exquisite refinement of cruelty; that they were, in many cases, starved to the bone, while flogged to their work, and that even in some instances, they were driven to commit suicide to evade the cruelties of a world, in which, though born so recently, their happiest moments had been passed in the garb and coercion of a workhouse."

"The profits of the manufacturers were enormous; but this only whetted the appetite that it should have satisfied, and therefore the manufacturers had recourse to an expedient that seemed to secure to them those profits without any possibility of limit; they began the practice of what is termed "*night working*," that is, having tired out one set of hands, by working them *through the day*, they had another set ready to go on working *through the night;* the day set getting into the beds that the night set had just quitted, and in their turn again, the night set getting into the beds that the day set quitted in the morning. It is a common tradition in Lancashire, that the beds *never got cold !*"

"These outrages on nature, nature herself took in hand; she would not tolerate this, and accordingly she stepped forth with an ominous and awful warning—contagious malignant fevers broke out, and began to spread their ravages around; neighborhoods became alarmed, correspondences appeared in the newspapers, and a feeling of general horror was excited when the atrocities committed in those remote glens became even partially known."

The above is copied from a work by John Fielden, Esq. M. P. for Oldham, and cotton manufacturer at Todmorden, in Lancashire.

My second extract is taken from the evidence of Sir Robert Peel, (father of the late prime minister,) as given before Parliament. Sir Robert is said to have had more parish apprentices than any one man in England. He thus speaks of them :

"Having other pursuits, it was not often in my power to visit factories, (speaking of his own,) but whenever such visits were made, I was struck with the uniform appearance of bad health, and, in many cases, stinted growth of the children. The hours of labor were regulated by the interests of the overseer, whose remuneration was regulated by the quantity of work done."

He further says:

"Such indiscriminate and unlimited employment of the poor, will be attended with effects to the rising generation so serious and alarming, that I cannot contemplate them without dismay; and thus that great effort of British ingenuity, whereby the machinery of our manufactures has been brought to such perfection, instead of being a blessing to the nation, will be converted into the bitterest curse."

It will be necessary here to state, that Sir Robert Peel, senior, introduced a Bill into Parliament in the year 1802; and thus commenced the factory legislation, which has been carried on with very little success, for more than forty-four years.

On the 3d of April, 1816, Mr. R. Gordon made the following statement in the House of Commons:

"It appears that overseers of parishes in London are in the habit of contracting with the manufacturers of the north for the disposal of their children; and these manufacturers agree to take one idiot for every nineteen sane children. In this manner wagon loads of these little creatures are sent down to be at the perfect disposal of their new masters."

I will take another extract from the evidence of L. Horner, Esq., one of the head Inspectors of factories. He says:

"These children were often sent one, two, or three hundred miles from the place of their birth, separated for life from all relations, and deprived of the aid which even in their destitute situation they might derive from friends."

He describes this as "repugnant to humanity, and a practice that had been suffered to exist by the negligence of the legislature."

In referring to the results of this inhuman practice, he says,

"It has been known that with a bankrupt's effects, a gang (if he might use the term) of these children had been *put up for sale*, and were advertised publicly as *a part of the property*. A most atrocious instance had come before the King's Bench, in which a number of these children, apprenticed by a parish in London to one manufacturer, had been transferred to another, and had been found by some benevolent persons in a state *of absolute famine*. Another case, more horrible, had come to his knowledge, while on a committee of the house; that an agreement had been made between a London parish and a Lancashire manufacturer, by which it was stipulated that with every *twenty sound* children, *one idiot* should be taken."

I will conclude this letter by making an extract from the speech of Sir Robert Peel, late prime minister, in Parliament. The speech does great credit to Sir Robert. He says:

"I am one of those who have derived our fortunes from the industry of the operative classes, and I trust that others who owe their prosperity to the same cause will feel as I do, that it is our *duty* to relieve the public, by taking on ourselves the charge of a just requital to those classes from whom our prosperity has sprung."

I have thought it necessary, in order to convince the reader of the truth of what I have stated, (the facts being of such a nature as to require the strongest testimony) to make the above extracts. I have taken them from the writings of men of the highest standing in society, and would be very willing to convince any one that I have not only copied them correctly, but selected them with an eye to mildness, rather than severity.

LETTER IX.

HISTORY OF AN ORPHAN BOY.

Dear Reader—Permit me to introduce to your notice an old acquaintance of mine, who, for the present, we will call CHARLES SMITH.

In my journeys through England as a traveller, it was my duty to make occasional calls at Manchester, the chief seat of the cotton manufacture; and on one of those occasions I was introduced by a mutual friend to the individual above mentioned. We became friends, and during our intercourse he related to me many of the incidents of his past life. The last time I had the pleasure of seeing him he put into my hands a parcel of documents relating to his history, with full permission to use them at my discretion. From these papers I select the following particulars.

Charles Smith has no recollection whatever of his parents; but from the documents before us, it appears he was born in the year 1792; and was removed to St. Pancras workhouse, in the north-western suburbs of London, in 1796. Being then about four years old, he said he perfectly recollected riding in a coach to the workhouse, accompanied by some female. He did not, however, think this female was his mother, for he had not the least consciousness of having felt either sorrow or uneasiness at being separated from her, as he naturally supposed he should, if she had been his mother. He thinks he had been nursed by his mother, but had passed through many hands before being taken to the workhouse, because he had no recollection of having experienced a mother's caresses. Young as he was, he often inquired of the nurses when the relations of other children came to see his young associates, *why no one came to see him*, and used to weep, when he was told, that *no one had ever owned him*, after his being placed in that house. It was supposed that he was an illegitimate child, that his father moved in the upper circle of society, and that his mother had died, probably from grief and disappointment, previous to his removal to the workhouse. Be this as it may, it is certain that when he applied, (after he arrived at manhood) to the parish officers of St. Pancras for information concerning his parents, they refused to give him any account of them.

The sad consciousness that he stood alone in the world, that he had no acknowledged claim of kindred with any human being, rich or poor, so constantly occupied his thoughts, that, together with his sufferings, they imprinted a pensive character on his features, which probably neither change of fortune, nor time itself, will ever entirely obliterate. He well remembers, when about six years old, as the children were repeating their catechism, it was his turn to repeat the fifth commandment; and as he was saying "Honor thy father and thy mother," &c., he burst into tears, and felt greatly distressed. Being asked why he cried, he innocently replied, "I cry because I cannot obey God's commandments; I know not either my father or my mother."

Smith acknowledges he was well fed, decently clad, and comfortably lodged, and not at all over-worked; yet with all these blessings, this destitute child grew melancholy. He relished none of the humble comforts he enjoyed. It was liberty he wanted. The busy world lay outside the workhouse gates, and those he was seldom permitted to pass. He was too young to understand the necessity of the restraint to which he was subjected. Like a bird newly caged, that flutters from side to side, and beats its little wings against its prison walls, in hope of obtaining its liberty; so young Smith, weary of confinement, and anxious to be free, often watched the gates of the house, in the vain hope that some favorable opportunity might facilitate his escape. He was so weary of confinement, he said he would gladly have exchanged situations with the poorest of the poor children, whom, from the upper windows of the workhouse, he had seen begging from door to door, or offering matches for sale to the people as they passed.

From this state of mind, Smith was suddenly diverted, by a rumor that a day was appointed, when the master chimney-sweepers of the metropolis were to come and select such a number of boys as apprentices till the age of twenty-one, as they might deem it necessary to take into their fraternity. These tidings sounded like music to the ears of Smith he anxiously inquired of the nurses if the news were true, and if so, what chance there was of his being one of the elect. The matrons told him that if he was elected he would bitterly rue the day that should consign him to that wretched employment. Still he was not satisfied that it could be worse than the state in which he then was.

The day arrived, the boys were brought forth; many of them in tears, and very sorrowful. Smith might be said to be the most joyous of the whole, and as the grim looking men approached him, he held his head as high as he could, and endeavored to attract their attention. Boy after boy was taken in preference to Smith, who was often handled, examined and rejected. Some of the sweeps complimented him for his spirit, and said, if he made a good use of his time, and contrived to grow a head taller, he might do for them the next time they came.

The confinement that was so wearisome to young Smith must have been equally irksome to his companions, therefore the love of liberty could not have been the sole cause of the difference of feeling manifested by these boys on this occasion. There was another reason; most of the boys had friends or relations, but poor Smith stood alone! No ties of consanguinity bound him to any particular portion of society, or to any place; he had no friend to soothe his troubled mind—no domestic circle to which, though excluded for a time, he might hope to be reunited. When the friends or relatives of other children came to visit them, the caresses that were sometimes exchanged, the joy that beamed on the faces of the favored ones, were any thing but pleasing to our young friend; not that he was envious of their happiness, but that it reminded him more forcibly of his forlorn condition.

From the period of Smith's disappointment in being rejected by the sweeps, a sudden calm succeeded, which lasted till another rumor was spread through the house, that a treaty was on foot between the Overseers of St. Pancras, and the owners of a large cotton factory, for the disposal of a great number of children. This occurred about a year after the chimney-sweep disappointment.

The rumor itself inspired Smith with new hopes; and when he found that it was not only confirmed, but that the number wanted was so great, that it would take off most of the children in the house, his joy became unbounded. Poor, infatuated boy! delighted with the hope of obtaining a greater degree of liberty, he dreamed not of the misery that impended, in the midst of which he afterwards looked back to St. Pancras as to an elysium.

Prior to the show day of the pauper children to the cotton manufacturer, the most illusive and artfully contrived stories were spread, to fill the minds of these poor infants with the most absurd and ridiculous errors, as to the real nature of the servitude to which they were to be consigned. From the statement of the victims to this bondage, it seems to have been a constant rule with those who had the disposal of parish children, before sending them off to the cotton mills, to fill their minds with the same delusion. Their hopes being thus excited, it was next stated to these innocent victims, that no one could be *compelled* to go, nor any but volunteers accepted.

When it was supposed that these excitements had operated to induce a ready acquiescence in the proposed migration, all the children, male and female, who were, or appeared to be, seven years old, were assembled in the committee room for the purpose of being examined, touching their *health*, *capacity* and *willingness* to go and serve as apprentices in the way and manner required, for the term of *fourteen years*.

The boys were to be instructed in *cotton spinning* and *stocking weaving;* the girls in *cotton spinning* and *lace making*. There was no specification whatever, as to the time their masters were to be allowed to work these poor children, although at *this period*, great cruelties were known to be exercised by the owners of cotton mills upon their apprentices.

Thus did the church-wardens and overseers of the poor of St. Pancras parish in the month of August, 1799, make over to Messrs. Lambert's, cotton spinners, hosiers and lace men, of St. Mary's parish, Nottingham, our young orphan boy, together with *seventy-nine* other boys and girls as parish apprentices, till they arrived at the age of twenty-one years.

The poor deluded young creatures were so inflated with joy that they began to treat their old nurses with insolence, and refused to associate with children, who, from sickness, or being under age, had not been accepted. But their illusion soon vanished, and they were soon made to endure hardships such as they had never conceived of.

Happy, no doubt, in the thought of transferring the burden of the further support of eighty young paupers to other parishes, the church-wardens and overseers distinguished the departure of this juvenile colony by acts of munificence. The children were completely new clothed; each had two suits, one for working, and another for their holiday dress. One shilling in money was given to each child, a new pocket handkerchief, and a large piece of gingerbread.

According to his own account, Smith was first to the gate. Having no relatives to take leave of, all his anxiety was to get outside. He was also the first to mount the wagon, and the loudest in his cheering. The whole convoy were well guarded by the parish beadles in their robes of office, and bearing staves, on their way to the wagons that were to carry them to their destination; but these officers the children were taught to consider as a *guard of honor*.

Some active young men were appointed to look after the passengers in the two large wagons, on their journey to Nottingham. Those vehicles were so secured that when once the grated doors were *locked*, no one could escape. Plenty of clean straw was put in for the children to sleep on, but they soon began throwing it over one another, and seemed delighted with the commencement of their journey. A few hours progress considerably damped this exultation. The inequality of the road, and the heavy jolts of the wagon, occasioned them many a bruise. Although it was in the middle of August, the children felt very uncomfortable in being thus cooped up in so small a space, (forty in each wagon,) and having no liberty except to look through the gratings of their prison, like so many wild animals on their way to an exhibition. After having passed one night in the wagon, many of the children began to repent, and express a wish to return. They were told to have a little patience till they arrived at Messrs. Lamberts, when no doubt those gentlemen would pay every attention to their wishes. Smith was so overjoyed with his prospects, that he spent *his shilling* at Leicester, in apples. The greater part of the children were much exhausted, and many of them seriously indisposed before they arrived at Nottingham.

After having been well refreshed, the whole of the children were drawn up in rows to be *reviewed by their masters*, their friends and neighbors. In Smith's estimation, the Messrs. Lamberts were "stately sort of men." They looked over the children, and finding them all right according to *invoice*, exhorted them to behave with proper humility and decorum; to pay the most prompt and submissive respect to the orders of those who would be appointed to instruct and superintend them at the *mills;* and to be diligent and careful, each one to execute his or her task, and thereby avoid the *punishment* and *disgrace* which awaited idleness, insolence, or disobedience.

This harangue, which was delivered to them in a severe and dictatorial tone, increased the apprehensions of the children, but not one durst open his mouth to complain. The masters talked to them of the various sorts of labor to which they were to apply themselves; but to the consternation of Smith and his associates, not the least allusion was made to the many fine things which had so positively been promised them while in London. This conversation seemed to look forward to close, protracted toil.

The children rested one night at Nottingham, in the *warehouse* of their new masters; the next day, in order to cheer their spirits a little, they were led out to see the local curiosities, which are so celebrated by bards of ancient times. The day following, they were conveyed in carts to the place that was to be their home for the next fourteen years. This place was Lowdham cotton mill, situated near a village of that name about ten miles from Nottingham, on the Surhill road. They arrived rather late in the evening.

The mill, a large and lofty edifice, being surmounted by a cupola, Smith at first mistook for a church, and expressed his opinion to that effect; this seemed to please some of the bystanders, who said he would soon know what sort of service was performed there. There was one source of consolation, he thought—it is not surrounded with large walls and strong gates, like the St. Pancras workhouse.

When the first cart, in which was young Smith, drove up to the door of the *apprentice-house*, which was half a mile distant from the mill, a number of villagers flocked round to see the young cocknies. One old woman said, "Eh! what a fine collection of children, see their pretty rosy cheeks." Another, shaking her head, said, "The roses will soon be out of bloom in the mill." "The Lord have mercy upon them," said a third. "They'll find no mercy here," said a fourth. In common with his comrades, Smith was greatly dismayed by these gloomy prognostications, which their guardians did all they could to check, or prevent the children hearing, hurrying them as rapidly as they could into the house.

The young strangers were conducted into a spacious

room, fitted up in the workhouse style, with long, narrow deal tables, and wood benches for seats on each side. The room seemed tolerably clean; but there was a rank, oily smell, which Smith did not much admire. They were ordered to sit down at these tables, the boys and girls apart. The other apprentices had not left work when these children arrived. The supper set before them consisted of milk-porridge, (i. e. oat-meal boiled in milk and water) and rye bread, very black and soft. Smith says it stuck to their teeth like bird-lime. As the young strangers gazed mournfully at each other, the governor and governess, as the master and mistress of the apprentices were styled, kept walking round them and making very coarse remarks. The governor was a huge, raw-boned man, who had served in the army, and had been a drill sergeant. Unexpectedly, he produced a large horsewhip, which he clanged in such a manner that it made the house re-echo. In a moment the children, who had been laughing and joking about the bread, &c., were reduced to the most solemn silence and submission. Even young Smith, who had been one of the ringleaders in these proceedings, was bereft of all his gaiety by the tremendous clang of the whip, and sat as demure as a truant scholar, just previous to his flogging; yet the master of the house had not uttered a single threat, nor, indeed, had he occasion, his stern and forbidding aspect, and his terrible horsewhip inspired quite as much terror as was requisite. Knowing that the apprentices from the mill were coming, this formidable being retired, to the great relief of the young strangers; but so deep an impression had he created, that they sat erect, scarcely daring to look on one side or the other.

While they were in this subdued state, their attention was suddenly attracted by the loud shouting of many voices; almost instantly the stone room filled with a multitude of young persons of both sexes, from young women down to mere children. Their presence was accompanied by a scent of no very agreeable nature, arising from the grease and dirt acquired in the avocation. The *boys*, generally speaking, had nothing on but a shirt and trowsers. Some few, and but a few, had jackets and hats. Their coarse shirts were open at the neck, and their hair looked as if a comb had seldom been applied. The *girls* were destitute of shoes and stockings, their locks pinned up, they wore no caps, few had gowns, the principal article of dress being long aprons with sleeves, made of coarse linen, that reached from the neck to the heels. Smith and his companions were almost terrified by the sight of the pale, lean, sallow looking multitude before them.

On their first entrance, some of the old apprentices took a view of the strangers; but the great bulk first looked after their supper, which consisted of boiled potatoes, distributed at a hatch-door, that opened into the common room from the kitchen. At a signal given, the apprentices rushed to this door, and each, as he received his or her portion, withdrew to the table.

Smith, who had always been accustomed to cleanliness, was much surprised to see the girls hold up their greasy aprons in which to receive their supper, and the boys resorted to acts even more filthy and indecent, when receiving the hot boiled potatoes allotted them. With a keen appetite the hungry apprentices devoured their supper, and seemed anxiously to look about for more. Next, the hungry crew ran to the table of the new comers, and voraciously devoured every crust of bread, and every drop of porridge they had left; after which they put and answered questions as occasion required. There was no cloth on the table, no plates, knives or forks; a little salt scattered here and there, with a plentiful supply of cold water, made up the usual accompaniments of the supper table.

The supper being devoured, in the midst of the gossiping that ensued, (for many of the old apprentices had come from the same parish, and were anxious to hear from their old nurses) the bell rang that gave the signal to go to bed.

The grim governor entered to take the charge of the newly arrived boys, and his wife, acting the same part by the girls, appeared every way suitable to so rough and unpolished a mate. She was a large, robust woman, remarkable for a rough, hoarse voice and ferocious aspect. In a surly, heart-chilling tone, she bade the girls follow her. Tremblingly the little creatures obeyed, scarcely daring to cast a look at their fellow travellers, or bid them good night. They separated in mournful silence, the tears trickling down their cheeks; not a sigh was heard, or a word of complaint uttered.

The room in which Smith and several of his companions were deposited was up two pair of stairs. The bed places were a sort of cribs, built in a double tier all round the chamber. The apprentices slept two in a bed. The beds were of flocks. From the quantity of oil imbibed in the apprentices' cloths, and the impurities that were suffered to accumulate from the cotton, a most disagreeable odor scented these rooms. The governor called the strangers to him and allotted to each his bed place and bed-fellow, not allowing any of the newly arrived inmates to sleep together.

The boy with whom Smith was to chum, got into his berth, and without saying a prayer or any thing else, was soon fast asleep. It was not so, however, with our young friend; he could not refrain from tears; he felt, young as he was, that he had been grossly deceived. When he crept into bed, the stench of the oily cloths and the greasy skin of his sleeping comrade almost turned his stomach. Over and over again the poor child repeated every prayer he had learned, and strove to recommend himself to the Father of the fatherless. At last, sleep did seal his weary eyelids; but short was the repose he was allowed to enjoy. Before 5 o'clock he was awaked by his bed-fellow, who springing upright at the loud tolling of the factory bell, told Smith to dress with all speed, or the governor would flog him. Before Smith had time to perform this office, the iron door of the chamber, creaking upon its hinges, was opened, and in came the terrific governor with the horsewhip in his hand; at the sight of him every boy hastily turned out of his crib, and huddled on his clothes in haste.

Smith and his fellow-travellers were slowest, not being rightly awake. He said to one of the boys, "Bless me, have you church service so soon?" "Church service, you fool," was the angry answer, "it is to the mill service you are called, and you had better look sharp or you'll catch it." The governor, bearing the emblem of arbitrary rule in his hand, walked around the chamber, looking in every bed place, amusing himself the while with cracking the huge whip, which fully understanding, the boys hastened below. Arrived there, they saw some of the boys washing themselves at a pump, and they were directed to do the same; after which, the whole multitude sat down to a breakfast at 5 o'clock in the morning. The meal consisted of a scanty supply of black bread and milk porridge.

They reached the mill about half past 5 o'clock; the machinery was going in all the rooms from top to bottom. Smith and his companions were astonished at the burring noise which proceeded from 20,000 wheels and spindles in motion, and many of them began to feel sickly from the rank smell of oil which bathed the machinery.

The new hands were received by Mr. Baker, the head manager, in a large room. They were then divided into separate divisions. The overlookers were then ordered to take the division each had assigned him. In this manner they were marched off to the different rooms, and immediately set to their work. All this was done amid the laughs, jokes, and coarse remarks of the overlookers, at the expense of the poor children.

The task first allotted to Smith was to pick up the loose cotton that fell upon the floor. Apparently nothing could be easier, and he set to with diligence, although much terrified by the whirling motion and noise of the machinery, and not a little affected by the dust and flue with which he was half suffocated. He soon felt sick, and by constant stooping, his back ached; he therefore *took the liberty to sit down;* but this attitude, he soon found was strictly forbidden in cotton mills. His task-master gave him to understand that he *must keep on his legs.* He did so till twelve o'clock, being six and a half hours without intermission. The poor boy suffered much from thirst and hunger. The moment the bell rang for dinner, all were in motion to get out as soon as possible. Never before did he enjoy the fresh air so much as now.

He had been sick almost to fainting, and it revived him instantly. The cocknies mingled together on their way, to communicate to each other their sad experience. When they arrived at the dinner room, each had his place assigned him.

During the first ten days Smith was kept picking up cotton; he felt at night very great weariness, pain in the back and ankles, and he heard similar complaints from his associates. After this he was *promoted* to the employment of a *roving-winder*, and being too short of stature to reach his work, he had to stand on a block of wood; but even with this help he could not keep pace with the machinery. In vain the poor child declared he could not move quicker. He was beaten by his overlooker with great severity, until, in a short time, his body was discolored by bruises. In common with his fellow apprentices, Smith was wholly dependent upon the mercy of the overlookers, whom he found to be, generally speaking, a set of brutal, illiterate men; void alike of understanding or humanity. Smith complained to the manager, who said, "do your work well, and you'll not be beaten."

It will be necessary to state here, that the overlookers had a certain quantity of work to perform in a given time. If every child did not perform its allotted task, the fault was imputed to the overlooker, and he was discharged. On the other hand a *per centage* was given to the overlooker, upon all work done more than the stipulated task. If, therefore, any complaint was made, the overlooker could have said, that if the owners insisted upon so much work being extracted from the apprentices, and a greater quantity of yarn produced than it was possible to effect by fair and moderate labor, *they must allow them severity of punishment*, to keep the children in a state of continual exertion. Each of the task-masters, in order to acquire favor and emolument, urged the poor children to the very utmost.

At the expiration of six months, being half starved, and cruelly treated by his task-master, Smith resolved to attempt an escape, to beg his way up to London, and lay his case before the officers of St. Pancras. In this attempt he could not get any of his companions to join him; he therefore determined to go alone. Steady to his purpose, he took the first favorable opportunity, and when the overlooker and manager were busy, he started off in his working clothes unperceived.

He began at a smart trot, looking behind him every

50 or 100 yards the first half mile; but finding he was not pursued, he slackened his pace. He continued to proceed as fast as he was able, not knowing whether he was on the right road to London, and being afraid to ask till he came to a village a few miles from the mill, where he was stopped by a man in the employ of his master. This person, it appears, had a commission from the mill owners, and according to agreement, was to receive five shillings for every runaway apprentice he caught and took back to the mill. Back poor Smith was taken, punished, and jeered at by his task-master, and worked with an increased severity. The poor children pitied him, but could afford him no assistance.

Their condolements, however, were grateful to his wounded pride and disappointed hopes. As he retired to his miserable bed, the governor, grinning, made him a low bow in the military style, and gave him a kick at the same instant. This afforded amusement to that portion of the elder apprentices who had made similar attempts and failed.

Many of these children had by this time been more or less injured by the machinery, several had the skin scraped off their knuckles, others had their fingers crushed, or taken off; young Smith, soon after his attempted escape, lost the fore finger of his left hand. One poor girl had been caught by the main shaft, and was so dreadfully maimed and mutilated as to be obliged to walk on crutches the remainder of her life, yet without having the power of getting any compensation or redress.

Many of the older boys were so oppressed with hunger, that they sallied out at night to plunder the fields; and declared their intention to do some crime, to get themselves transported, in order to be freed from their cruel task-masters.

When Smith had been four years with his masters,

they gave up the business, from what cause is not known to me.

There was at this time a mill owner in one of the villages of Derbyshire, of the name of Needham. Like many of his class, he had risen from a state of abject poverty, and had it been by honorable industry, his prosperous fortune would have redounded to his credit. Of his original state of poverty it was his weakness to be ashamed. By the profusion of his table, and the splendor and frequency of his entertainments, he seemed to wish to cover and conceal his mean descent. His house, lawns, equipage, and style of living, completely eclipsed the neighboring gentry; yet, boundless as was his ostentation, he was in his heart sordidly mean; which was sufficiently evinced by his cruelty, in wringing from poor, friendless orphans, the means of supporting his unbecoming pomp. His mansion was at Highgate Wall, near Buxton.

To this unfeeling master, Messrs. Lamberts made over the unexpired term of years, for which the parish apprentices had been bound by their respective indentures. What premium was paid, or if any, I know not; but as this man was neither a hosier, nor a lace manufacturer, he had not the power to fulfil the conditions, of instructing the children in lace-knitting and stocking weaving. The consequence was that they lost the most important advantages of their servitude; and those who survived their term of apprenticeship, found themselves without that degree of skill which was requisite to enable them to gain their bread.

Mr. Needham went to Loudham to inspect the children, and was very liberal in his promises of kind treatment. The whole lot, male and female, to thc amount of many scores, were then removed in carts from Lowdham to Litton mill. The first day's progress brought them to Cromford, where they halted for the night. The girls were lodged in dwelling houses, the boys slept on straw in a barn and stable ! The next morning the whole party were marched on foot through the village. Then they again mounted their carts. It was in the month of November when this removal took place. On the evening of the second day's journey, the devoted children reached the mill.

This was situated at the bottom of a sequestered glen, surrounded by rugged rocks, and remote from any human habitation. In this place where our young friend spent the next ten years of his life, many of these poor children were hurried to an untimely grave.

The appearance of the apprentices who were at work in the mill when the new hands arrived, was any thing but pleasing. The pallid, sickly complexions, the filthy and ragged condition of their clothing, gave a sorrowful foretaste of what apparently awaited him. From the mill they were escorted to the apprentice house, where every thing wore a discouraging aspect. The lodging room, the bedding, &c., all betokened a want of cleanliness and comfort.

Smith passed a restless night, bitterly deploring his hard destiny, and trembling at the thought of greater sufferings. Soon after four o'clock in the morning, they were summoned to work by the ringing of a bell. The breakfast hour was eight o'clock; but the machinery did not stop, they got it as best they could, now a bit and then a sup, all the time doing their work.

Forty minutes were allowed for dinner, a part of this time being required for cleaning the machinery. The number of the working hours at this mill was from fourteen to sixteen per day. The children suffered severely from this unnatural state of things. From all these sources of sickness and disease, no one will be surprised to hear that contagious fevers arose in the mill, nor that the deaths were so frequent as to require constant supplies of parish children to fill up the vacancies. It has been known that forty boys were sick at one time, being about one fifth of the whole number. Smith was one of the sick, and well remembers tar, pitch and tobacco being burned in the room, and vinegar being sprinkled on the bed and floor. He also remembers the doctor saying, "It is not drugs, but kitchen physic they want." So great was the mortality at one time that the mill owner deemed it necessary to divide the burials, sending a part of them to a distant village, although the fees were greater, in order to conceal the sad reality from his neighbors.

Not a spark of pity was shown to the sick of either sex; they were worked to the very last moment it was possible for them to work, and when they were no longer able to stand, they were put into a wheelbarrow, and wheeled to the apprentice house. The doctor was seldom called, till the patient was in the agonies of death.

I would not willingly overcharge the picture I am drawing, or act so unwisely as to exaggerate these atrocities; and it is with some degree of diffidence I state, in consequence of combined and positive testimony, that no nurse, or nursing, was allowed to the sick, further than what one invalid could do for another!—that neither candle nor lamp-light was allowed, or the least sign of sympathy manifested.

I will not harrow up the feelings of my readers, by entering into a minute detail of all the hardships and sufferings that befel poor Smith during his fourteen years servitude in these places; they are such as would scarcely be credited in this land. It will be sufficient to say, that in addition to his attempt to escape, he twice run off to make complaints to the magistrates, and show them his bruised and crippled frame, but without obtaining the least relief. Finding no remedy in the law, he seriously entertained the idea of committing suicide; for this purpose he went up to an attic window in order to throw himself out. While contemplating the awful leap he was about to make, something seemed to whisper, "Have faith, and struggle on." He then abandoned the idea.

Having completed his term of fourteen years, viz., from seven till twenty-one, he worked till he had saved a little money, and then left that place, which had been the scene of so much suffering to him and his companions. Many of the children who left St. Pancras with him, had gone the way of all flesh, one had become an idiot, in consequence of ill treatment; one girl was unable to walk without the aid of crutches; and the remainder were more or less crippled and mutilated in various parts of their persons. Smith was a sad spectacle to look upon, stunted in growth, his legs twisted by standing so many hours at the frames, his countenance haggard and care-worn; altogether he presented the appearance of a man of seventy years, who had seen much service, rather than a person of twenty-one years, just entering upon manhood.

Although he had been thus crippled, and had not been taught the business to which he had been bound, yet there was *no law in England* whereby he, in his poor condition, could obtain compensation for the injuries he had suffered, or might suffer through life, in consequence of the unfeeling avarice of his masters.

The subsequent history of Charles Smith is one continued series of trials, arising out of his crippled condition; and though it might be interesting to the general reader to know how, by patient, persevering industry, he afterwards so far overcame the difficulties of his position as to become a small tradesman in Manchester, yet these particulars would far exceed our limits.

When I last saw Smith in Manchester, he told me he had a short time before been *burned out;* but that he was then beginning to recover from his losses. He had a wife and three children, two girls and a boy; and while I took tea with him, he told me he would feel quite happy if God would enable him to keep his children from going into the factories.

It gives me great pleasure to say that he is looked upon in his neighborhood, as an honest, industrious man, a good husband and kind father.

LETTER X.

THE LACE MAKERS OF NOTTINGHAM.

Lace making is a business in which men, women and children, are promiscuously engaged. This business requires in its various operations, the delicate little fingers of children, perhaps to a greater degree than any other. This is one reason why a large majority of the infant laboring population of Nottingham are employed in lace manufacture. Lace is chiefly fabricated in machines driven by steam. These machines are very complicated in their structure, but as my object is more particularly with the human beings who attend them, I will refer the curious reader to the various works upon this subject, in many of which a full description of those machines is given.

It will, however, be necessary to state, that the machine in which the lace is made, is supplied with the *thread* by means of small *bobbins.* The process of filling these bobbins with thread, is called *winding*, which is performed by young women of from fourteen to twenty years of age; they very rarely begin this branch earlier, as it requires *great care*. The bobbins when filled with thread, pass into other hands, to be placed in a carriage, at the top of which is a hole, not larger than the *eye of a needle*, through which the thread must be put. This operation, called *threading*, is chiefly performed by boys. The bobbins, with their carriages, supply the machine with thread; somewhat similar to the "shuttle" in common weaving. After the bobbins are placed in the frame, it is the duty of a workman to superintend their motions. He has to watch the whole breadth of a machine, weaving a piece of lace perhaps forty, fifty, or sixty inches wide; and in which 3600 bobbins pass through as many guide-threads, a hundred times a minute. This may be thought impossible; it is nevertheless true. Should any fault occur he must adjust it on the instant. The lace machine is frequently kept going for eighteen or twenty hours out of the twenty-four.

When the piece of lace in the frame is finished, it is taken out; the *winders* and *threaders* are then required to fill the machine with the requisite number of bobbins to commence a new piece. When this takes place in the night time, as is very frequently the case, and all this threading has to be done in the glare of a gas light, it is very distressing for the eyes. The poor children require sometimes to be *shaken*, or *beaten with a cane*, to keep them from falling asleep from excessive fatigue. They are mostly divided into night and day sets, and take their turns for each alternately. Though the work itself is not hard, the children are much harassed by night work. and irregular attendance.

After the piece of lace is taken from the frame, it undergoes a variety of processes, such as drawing, rectifying, embroidering, pearling and hemming. The article now leaves the factory, and is intrusted to small masters and mistresses, who work at their own houses, and employ children to perform the above operations. A large number of children, mostly girls, are employed in all these processes, which more or less are performed with the needle. In fact, I believe that almost all the children of the laboring classes in Nottingham, are engaged in one or the other of the several branches of the lace manufacture, and at a very early age. The common saying is, "as soon as they can tie a knot, or use a needle." It is in these departments of the trade that we see infant labor in its worst light. The number of hours these infant victims are kept incessantly at work, in confined apartments, and the tender ages at which they are put to it, would be almost incredible to a stranger. All this, however, is substantiated by facts, which place the matter beyond the possibility of a doubt. Almost all the families employed in the lace manufacture of Nottingham, are supported, more or less, by the labor of their children.

One of the great evils of this system, is, that of *reversing* the order of nature; children become at an early age independent of their parents,—in many cases the latter are even obliged to act as menials to their children. Another evil, is, that worthless fathers are enabled to spend their time in low pot-houses, out of their children's earnings. A third evil is the immorality which prevails among the young people. The *threaders*, who are usually boys, and the *winders*, who are mostly girls, are required at the same time and place, day and night; and thus, in the absence of proper restraint, have every facility for forming improper connections. The natural results of such a noxious system are but too apparent, and must have contributed in no slight degree to the immorality, which, according to the opinion universally expressed, prevails to a most awful extent in Nottingham.

Women brought up under such a system know little of the domestic duties of every day life, and which are so essentially necessary to be known by those who may be called upon to fill the important stations of wives and mothers. Hence we find them after marriage still engaged at their own houses in some process of lace manufacture, while the duties of the family are intrusted to others.

Having no time to attend to their families, nor even to discharge the first and most sacred duty of mothers, that of nursing their offspring, they freely administer opium in some form, such as Godfrey's and Anodyne cordial, and laudanum, to their infants. These drugs are given to infants at the breast, not because the child is ill, but to compose it to rest, in order to prevent their cries interfering with the protracted labor by which they strive to obtain a miserable subsistence. The infants become pale, tremulous, and emaciated, the joints and head enlarge, they become listless, and death at length steps in to their relief.

Great numbers of children are thus carried off yearly; should they, however, get over "the seasoning," as it is called, they begin to come round about three or four years old, i. e., as soon as the laudanum is discontinued.

In the present state of trade, it would be impossible for men to do without their wives laboring; *they must work*, however many children they may have; from the same cause, the children must go out to work as soon as they are able to use the needle.

With respect to wages, I will not venture to make any statement; I fear it would not be credited. This may be imagined from two things: first, the condition of the people; second, from the fact that a piece of lace which formerly was worth from seventy to eighty dollars, can now be bought for *three* dollars.

Such is a brief outline of the condition of one of the most industrious classes to be found in England. The fair wearers of lace will be distressed to learn that this highly ornamented article, is produced (in England, at least,) at the expense of so much misery.*

LETTER XI.

FLAX SPINNERS.

The condition of the operatives in flax mills, in Leeds, and other places in the North of England, and South of Scotland, is any thing but pleasing to contemplate. This will be best understood by inspecting the various processes, which are of the following nature.

The flax, as imported, is delivered to the *hand-hecklers*, who roughly separate the fibres by drawing the bunches through a quantity of iron spikes called heckles, fixed before them. This is a very dusty process. The hand-hecklers are mostly men, or strong lads. The flax is then carried to the *heckling machine*, in connection with which a greater number of young children are employed

* In 1846, a bill was brought into the House of Commons (prepared by Mr. T. Duncombe, Colonel Rolleston, and Mr. J. Fielden,) "to regulate the hours of night labor in factories where bobbin, net and warp lace machinery is employed," consisting of seventeen clauses. It proposes to enact that night labor shall henceforth cease in these factories, and that the working hours shall not be earlier than six o'clock in the morning to not later than ten o'clock at night, subject to certain penalties. It further prohibits the employment of children under eight years of age.

than in any other process in the business. They are mostly boys. The machine consists of various sorts of cylinders, or rather polygonal prisms, having heckles set on their edges, which revolve with great rapidity; and the business of the *machine minders* is to fix the bunches of flax on supports in front of these heckles, and to move them from time to time, from the *coarser* to the *finer* heckles. The bunches, for the purpose of being thus suspended, are screwed between two bars of iron, which is the business of the *screwer;* who is generally a *younger boy* than the machine minder, and his labor is very fatiguing; in fact, this is one of the most *laborious* employments to which children can be put, independently of the noxious atmosphere, which is loaded with particles of flax incessantly pulled off, and scattered by the whirling of the machines. *The screwer seems not to have an instant's cessation from labor;* bunch after bunch is thrown down before him to fix and unfix, which he performs with astonishing rapidity. If he does not perform his work properly, it mars the work of the *machine minder*, and a box on the ear, or a kick with the foot, is the usual consequence.

The machine minder is far from being idle; he has to move his flax when it has received its due proportion of heckling in one position, the arrival of which time is indicated by a bell; he has also to collect from between the rows of spikes, *as they revolve*, the tow, or short fibres and refuse of the flax which they comb off. The boys become very expert at this part of the business, but sometimes suffer severely while learning, in consequence of having their hands caught by the spikes on the cylinders as they revolve. The tow is collected and carried to the card room, which is equally bad in regard to dust as the heckling room. After the tow has been carded, it is ready for spinning. I believe that tow is invariably spun *dry*.

The heckled flax, or *line*, after being separated from the tow, is sorted, according to its fibre, for various degrees of fineness. This is done by young men called *line-sorters*. Girls termed *line-spreaders*, are employed to unite the bunches of line into one sliver, and thence it is roved and spun.

In spinning the fine line, it is necessary to allow each thread to pass through a *trough of hot water*, (from 110 to 140 degrees of Fahrenheit,) which is placed at the back of the spindles. This is called *wet spinning*. The hot water enables them to spin the line much finer than it could be done without it, on account of the fibres sliding more easily among each other. As the line is spun it is wound on to the spindles, and as they revolve very rapidly, they throw off a continual sprinkling of water along the whole front of the frame. Now as there is another frame at no great distance, the spinners, who are mostly young women, are exposed to this small rain both before and behind; which is quite sufficient to wet them through in a few hours, especially when the frames are placed too close to each other. They stand on a *wet floor*, which is so constructed as to let the water run off into pipes below, and thence into a common sewer. They have no protection from the hot water except a *blanket-apron*, which is soon wet through, and they generally work without stockings and shoes. Most of the time their hands are dabbling in the hot water, in piecing broken threads, and rectifying any error of the machinery.

I have seen these rooms so filled with steam from the hot water, (a fresh supply of which is constantly running into the troughs,) in the depth of winter, as to oblige them to keep the windows open to let it out.

The consequences of all this are frequent colds, occasioned by passing from so warm a room filled with steam, to the open air at all seasons, especially in winter evenings; hands much chapped and sore, which it is painful to behold, and considerable swellings of the feet and legs. It is no uncommon thing to hear of eight or ten of them remaining at home sick. Their appearance as they walk along the street, is that of persons far advanced in a decline.

The dry spinners suffer from the dust and small particles of flax, which get on to their lungs and cause asthma, &c. Many of the young women employed in wet and dry flax spinning, die early in life. Should they, however, live till thirty-five or forty years of age, they appear to have all the symptoms of old age.

Since I commenced this article, the "Morning Post" of April 3, 1846, has been handed to me, in which I find the following case of hardship experienced by six young women employed in the flax mills of Dundee, in Scotland, to which I would draw the reader's attention.

It appears that Messrs. Baxter of Dundee, (who are said to be wealthy and powerful persons, influential merchants, flax spinners, bankers and ship owners,) employ 2,500 persons, of whom 1,300 were employed in the factory in which the case to which we refer occurred.

The poor girls are six in number, the eldest nineteen, and the youngest thirteen years of age. Four of the six are orphans, entirely unprotected. They had worked for Messrs. Baxter a long time, and during the whole period of their service had never been guilty of any offence whatever. It appears that their respective wages were $1,32 a week, and that, as some of the operatives employed in the same mill with themselves had lately applied for, and obtained a rise of six cents, (i. e. one cent a day,) in their wages, viz ; from $1,32 to $1,38, they applied for a similar advance, and were *refused.* The rules of the factory obliged them to be at their work at 5 o'clock in the morning, and continue, with the exception of meal times, till seven o'clock in the evening. Also, that for all the time they were absent on their own account, the operatives were to pay a fine equal to "the time and a half;" i. e. if a person was absent five hours, he should be fined the amount he could earn in seven and a half hours. This is the *standing rule* in that mill.

On the 27th of October, 1845, these six girls, after leaving work at the dinner hour, did not return to it again that day, and were absent about five hours, viz : from 2 till 7 o'clock. On the following morning they were at work at 5 o'clock as usual, (not knowing that they had done any thing wrong, any further than breaking the rules of the mill, for which, they were aware, they would be fined to the amount of seven and a half hours' work,) and continued to work till 5 o'clock in the evening; when they were *apprehended while at work*, by four men, and taken to a private office and examined. They there stated that "They had no desire to desert their service, but had merely taken that afternoon for recreation."

They were kept at this office *five hours*, and were afterwards carried to another private office, where the magistrates were, together with one of their employers with his overseer or manager. They were then told they must sign a paper which was placed before them, and it is worthy of remark that Mr. Baxter was observed to whisper something to the magistrates before the *sentence* was passed upon them. And what does the reader think that sentence was? Why, nothing less than *ten days' imprisonment, with hard labor.* This mock trial was conducted with *closed doors*, no one was allowed to enter, even the few relatives of these poor girls were *refused* admittance. Their relatives were also refused to see them during their confinement. This is a specimen of the wretched laws of England. There is no justice for the poor.

This case was taken into consideration by the inhabitants of Dundee, who sent a petition to Parliament on the 26th of February, 1846.

It was again brought before the notice of Parliament on Thursday, April 2, when a motion was made to inquire into the particulars of the case, and *lost* by a majority of *twenty-five*. Alas ! poor country, when will thy oppressed people have justice done them?

There are some people who take a sort of pleasure in speaking against the laboring portion of my countrymen; calling them idle, drunken, and the like. In order to show the reader that such is not their general character, and that they are not only industrious, and work hard for a little, but that they know how to take care of that little when they have got it, I will here insert an extract, taken from the books of the *Savings Bank* of Dundee some four or five years ago.

In that town, out of 464 male weavers, with wages averaging $1,92, 108 are depositors. Of 181 flax dressers, with the wages averaging $2,88, 36 are depositors. Of 290 mechanics, with wages equal to $4,85, 56 are depositors. And even among the class of girls whose case I have just recorded, there is *one solitary* depositer, while there are 237 accounts in the names of female domestic servants.

The reader will bear in mind that *each* of the poor people whose hardships have been recorded, in this and the preceding letters, *have to pay taxes to government*, in addition to finding their own food, clothing, and necessaries, without regard to age, sex, or condition.

LETTER XII.

A FACTORY CONDUCTED ON CHRISTIAN PRINCIPLES.

In the town of Bradford, in the county of Yorkshire, England, there is one of the largest *worsted spinning* establishments in England; by some it is thought to be the largest in the world. It is carried on under the firm of Messrs. Wood and Walker. From actual observation, (having on several occasions been permitted to inspect the premises,) the writer is enabled to make the following remarks concerning this vast establishment. The reader's attention is called to this, as an *exception* to the general rule of factory government in England.

On the premises, which are very extensive, are two large mills, counting room, school house, &c. At the entrance gate is a porter's lodge, with a person constantly in attendance; no one being permitted to enter except by leave, or on business.

A person on entering this place for the first time, is struck with the clean, orderly, and healthy appearance of the work people; and is naturally led to inquire how these desirable results are brought about. The answer to these inquiries is to be fpund in the *mild*, *generous* and *christian* regulations of the place.

The hours of labor in these mills are *eleven* per day. In summer they commence work at 6 o'clock in the morning, breakfast at half-past 7, (30 minutes being allowed); dine at 12 (one hour being allowed); tea at 5 (30 minutes being allowed); and leave the mills at 7 in the evening. In winter they commence at half-past 6, deducting 10 minutes from each of the three meal times to make up the difference. No part of the time allowed for meals is occupied in cleaning machinery, but is entirely at the disposal of the work people.

Those work people who live at a distance, and who are obliged to bring their food with them, have a warm, convenient room provided for them to sit in, and their victuals are made warm and comfortable, as if they had been at home; this arrangement is attended with very little cost, while it enables every one to have warm, clean and comfortable meals.

The dining room is a large one, on the ground floor, kept for this, and no other purpose. It is provided with good strong tables and forms, and kept very clean and orderly. Adjoining this room is a smaller one, used for the sole purpose of warming the provisions of the work people; it is well fitted up with a steam apparatus, troughs, shelves, &c. The children and others, who live at a distance, being their breakfast, dinner and tea when they come to work, in *tin cans* (which are all numbered), and place them on the particular shelf allotted to the room in which they work. A man (and sometimes also a woman) looks after this room, and gets every tin made warm by means of the steam apparatus, and all placed on their respective tables in proper time. It gave me great pleasure to inspect the arrangements of these well regulated rooms.

In one of my visits, Mr. Balme, the schoolmaster, kindly accompanied me to the dining room; we took our station at the lower end of the room, directly opposite the entrance, and awaited the coming of the children. This, I was told, is part of the schoolmaster's duty, and his presence preserves silence and order during meal hours. Exactly at the half hour the engine stopped, and the children, to the number of about 400, began to come in to breakfast. All the tin cans containing their tea or coffee had been placed on the table ready, and all had taken their seats; the boys on one side of the room and the girls on the other; and had unfolded their little portions of bread and butter, but no one began to eat. How is this? I turned my eye to the schoolmaster for an explanation. He had his watch in hand, and at the expiration of 5 minutes from the time the engine stopped, (which time is allowed for all to get seated) he gave a signal. At this signal they all rose and sang a beautiful verse as *grace before meat;* this surprised and pleased me much; I could scarcely believe I was among factory people.

The grace being ended, they began to eat their breakfast, the schoolmaster still remaining in his place. Some of the boys were soon done, and seemed to manifest a little impatience to get out to play; these boys kept their eye steadfastly fixed upon the schoolmaster. At exactly 10 minutes from the time of commencing breakfast, he gave the word "go;" immediately the boys and younger girls departed quietly, two or three together, while the young women remained to spend their quarter of an hour in knitting or sewing.

Now, dear reader, look into the play-ground; see their merry little faces and active limbs, striving who can be most happy. How very different is all this from what I experienced when a factory boy. But how does it happen that these children are so active and playful, and the generality of factory children so jaded and tired? It is because these children enjoy many privileges, of which ninety-nine out of every hundred factory children know nothing.

In every room in which the children work, there are at least half a dozen spare hands, who relieve the others by turns, and the children are not only allowed, but provided with seats. These seats are fixed along both sides of the rooms, in addition to which, every frame attended by any young person has stools attached to it by means of a joint, which allows of their being placed under the frame when not wanted. The seats at the sides of the rooms are for the use of the spare hands; those attached to the frames, for the children at work. Occasionally it happens in these mills, that a child can have half a minute to spare, then by a motion of the foot, the stool can be brought from under the frame, and it can be thus relieved from its standing posture. None but those who have experienced it, can know the value of a seat in these spare half minutes, to a child thus circumstanced. The children have here no harsh treatment to endure from their overlookers, who seem to be an intelligent set of men, and endowed with a large share of the christian spirit of their masters. Should one of these overlookers dare to strike a child, he would be immediately dismissed.

The school-room is a large new building, erected near the mills. The firm provide, at considerable expense, a schoolmaster and schoolmistress, who are brother and sister; also books, slates, maps, pens, ink, &c. The children under thirteen years of age, to the number of between 300 and 400, are divided into sections, each attending school at least two hours per day. The boys learn reading, writing, arithmetic, geography and singing; and the girls learn knitting and sewing, in addition to the above. Adjoining the school-room is the washing-room, which is provided with a number of large basins and clean towels, and water is laid on and can be had by turning a tap. A quarter of an hour suffices to enable a division to clean for school. With a more cleanly, healthy-looking set of factory children I have nowhere met.

It is but justice to state, that this school and its arrangements were originally planned and superintended by that indefatigable friend of the factory laborers, the Rev. G. S. Bull; and that it was opened for the children previously to the introduction of the Factory Regulation Bill.

When the children arrive at the age of thirteen years, they are then permitted to work full time; and on leaving school they are presented with a handsome Bible, with the following inscription on the inside of the cover:—

"This copy of God's Holy Word was given to on attaining the thirteenth year of her age, as a reward for good conduct during three years attendance at Messrs. Wood and Walker's Factory School.

"May you ever 'read, mark, learn, and inwardly digest;' may you 'embrace and ever hold fast the blessed promises of everlasting life,' contained in this Sacred Book.

"May it be your guide through life, and your support in the hour of death.

M. Balme, Schoolmaster to
Messrs. Wood and Walker.

Bradford, 184—."

The young women in this establishment are of a superior cast. This arises in part from the care which has been taken of them when they were children, and from the rules respecting their government now they are grown up, which rules are strictly enforced. The principal of these are, that *no married females shall be allowed to work in these mills;* and that any "single female" being known to *conduct herself improperly, must instantly quit her employment.* The hours of labor also, not being so long as at most places, allow them more time to learn domestic habits and improve their minds. They enjoy also a great advantage in having warm, comfortable meals on the premises, if they should require it. Their appearance is clean and decent, and they seem to take a pride in keeping themselves so. On inquiry, I learned that most of them had been brought up under this firm, and may be said to know very little of the vice and wickedness generally prevailing in other English factories.

There are not many men employed in the spinning departments, but of wool-sorters and combers there are a great number, who enjoy comparatively good wages.

A surgeon, liberally paid, is provided by the firm; whose duty it is to go over the works *daily*, for the purpose of inspecting the health of the work people. Should this gentleman notice any one looking ill, he makes inquiry as to the cause, and if it be any thing requiring rest or medicine, he is ordered home immediately; or should any of the work people feel themselves sick, they apply to the surgeon and obtain timely advice and assistance. During the time they are off work, their wages are sent to them the same as if they were at work. To this part of their benevolent plan, the work people contribute a small weekly sum.

In a conversation with one of the partners of the firm, he said to me, "that, although they did all they could to make their work people comfortable, yet they were well aware *their* system was not what it ought to be. They were anxious to reduce the hours of labor to ten per day, if the other manufacturers would do the same; and until that took place, he did not see what other improvements could be made."

This establishment is conducted in a manner which reflects great credit on the firm, and affords a striking proof of what may be done by those manufacturers who feel disposed to improve the condition of their work people. It would give me great pleasure to be able to say, that of the many thousand establishments there are in England for the manufacture of silk, cotton, linen, and woollen goods, *five* could be found to answer the description of the one here given; such, however, I fear is not the case.

LETTER XIII.

THE CONTRAST.

I am aware that it is not always advisable to make comparisons; there are, however, exceptions to most rules, and if any good is likely to result, I think we ought not to allow any trifling circumstance to check our endeavors. In laying before the reader the following *contrast*, I am actuated solely by a desire to show the tendency of insufficient remuneration for labor, and the evil resulting from long hours of labor to young people in factories.

In the winter of 1841-2 my business required me to travel through the county of Derbyshire, in England. In this journey I was detained by unfavorable weather at Matlock Baths, a place remarkable for its romantic beauty, mineral springs, and subterranean caverns. I employed a part of the leisure time thus afforded me, in examining these local curiosities; but particularly the caverns, which are so justly celebrated.

Having satisfied my curiosity with all that was interesting under ground, I turned my attention to what was going on among the inhabitants of the village.

In Matlock Baths and the neighboring village of Cromford, there are three cotton mills. These mills, at the time of my visit, belonged to the late Mr. Richard Arkwright, the only son of Sir Richard Arkwright, the father of the English factory system.

As this place was one of the earliest seats of the cotton manufacture, I was anxious to see the effects of the factory system here. The people of this village seldom migrate, have very little intercourse in any way with the inhabitants of large towns, and know but little of what is going on beyond the beautiful hills by which they are surrounded.

I found the general condition of the people to be any thing but favorable to a high state of moral and intellectual culture. Long hours of labor, low wages, and hard fare seemed to be the prevailing characteristic of the factory system in that place. Many of the poor inhabitants related to me their tales of sufferings and privation, and seemed to feel their miseries very keenly. Among other things, I was very much interested in the case of one of their factory cripples.

Being directed to a small cottage in the village of Cromford, (which joins that of Matlock,) where this poor man resided, I made free to knock. The door was opened by a clean, elderly woman, the widowed mother of the poor cripple, who kindly invited me in, and requested me to be seated.

Having explained the object of my visit, I was directed to a young man who sat in a corner, at work upon a child's *first shoe.* After a short introduction, I asked him if he would be kind enough to inform me how he became such a cripple; he very readily complied with my request, and related to me his history in nearly the following words, his mother sitting beside him at the time:

"My name is J—— R——; I went to work in the cotton factory in the adjoining village of Matlock Baths, at the age of 9 years. I was then a fine, strong, healthy lad, and straight in every limb. They gave me at first 48 cents per week. For this I had to work 12 hours a day. Our master was very exact as to time; if we worked more than 12 hours a day we were paid for the extra time, if less, we were abated.

Sometimes we were obliged to stop the mill, from having too much or too little water; the time thus lost was always deducted, and our overtime added. On pay-day we gave in our time, and were paid accordingly."

"Then," said I, "you mean to say, that your wages of 48 cents was for 72 hours' work. To which he replied, "exactly so."

"I got my wages raised," he continued, "as I learned my business; but our master, Mr. Arkwright, was always very exact in these things. He always raised our wages at the beginning of the year, and at no other time except for some very particular reason. He would then give us a little more each, and that was to be our wages for the next twelve months. I worked in that mill for about ten years. I never worked in any other mill; I never had any other master than Mr. Arkwright.

I gradually became a cripple, till at the age of 19 I was unable to stand at the machine through the day, and was obliged to give it up. You see, sir, what a figure I am; I cannot walk without the help of this stick and my brother's arm. I have only been down to the market place, at the bottom of the street, twice since I left the mill, and I do not feel a desire to go again, the people stare at me so."

"How long is it since you left the mill?" said I.

"Nearly two years."

"Then you are about 21 years old?"

"Yes sir, I shall be 21 next birth-day."

Reader, imagine you see before you a young man, whose body forms (when he is standing, supported by a stick on one side and is holding by a table on the other) a curve from the forehead to the knees, similar to the letter C, his legs twisted in all manner of ways by standing at the frames, and you will have a tolerable picture of our friend J—— R——, of Cromford.

Here is a person just entering upon manhood, who was evidently intended by nature for a stout, able bodied man, crippled in the prime of life, and all his earthly prospects gone. Such a cripple as this man, I have seldom met with. Yet it was pleasing to see with what patience and resignation he bore his lot. He had learned to read and write a little, and his brother was teaching him to make first shoes for children.

This man has been paying taxes to the government of England from his birth, or his parents for him; but there is no law by means of which he can gain any compensation for the injuries he has sustained.

In returning to my lodgings, I passed by his master's CASTLE; and my imagination was busy at work in picturing the multitudes of human bones and sinews that had been sacrificed in building it. I almost fancied I could see them intermixed with the stones and mortar.

I left Matlock Baths, and in the noise and bustle of every day life, the case of J. R. had been well nigh forgot, till in the spring of 1843, it was again brought to my recollection, by seeing in the newspapers an account of the death of his former master, Richard Arkwright, Esq., of Wilersley Castle, near Cromford, Derbyshire.

This gentleman succeeded to all the possessions and numerous spinning factories of Sir Richard, his father, in 1792, then estimated at the value, capital stock included, of about *two million five hundred thousand* dollars. By his extensive spinneries in Cromford, Bakewell, and Manchester, he is said to have derived a clear income of 500,000 dollars annually. The extensive works at Manchester he disposed of sometime after his father's death, to his managers, Messrs. Barton and Simpson, who both realized large fortunes. In about 1837—8, he disposed of his spinning works at Bakewell; but those at Cromford, near his own residence, he carried on to the time of his death.

Mr. Arkwright died possessed of not less than thirty-five million dollars! in personal property alone, irrespective of landed estates.

As an individual capitalist, there is not one in Europe at the present time who can approach within half the distance, excepting, perhaps, the excellent, no less than wealthy Mr. Solomon Heine, of Hamburg; who, according to general repute, is estimated to concentrate in his own person the representation of money values to the amount of $20,000,000. It must be remembered, however, that this sum represents the whole property of Mr. Heine, whereas, the late Mr Arkwright was possessed of landed estates to the value of seven or eight millions of dollars, beyond the amount at which the personality is rated.

Immensely wealthy as are the Barings, the Rothschilds, the Hopes, &c., of Europe, there is not, has not been *one*, that could be placed at all in the comparison; not all the magnificent fortunes drawn off, with all the vast capital remaining still in the princely house of Baring, would reach to the amount; not all the capital of all the Rothschilds throughout Europe together, would equal more than one half the enormous mass of wealth left behind by the late Mr. Arkwright.

The reader will bear in mind that Mr. Arkwright was the son of Sir Richard Arkwright, who, about the year 1778 was a *common barber* in Preston, Lancashire, shaving for *one cent* per head. On one occasion the Preston barber had to make his appearance in a court of justice;

but his clothing was so mean that he was ashamed to go, and being too poor to purchase any thing new, his friends and customers entered into a subscription of one shilling each, to put him in decent plight.

How many thousands of human beings have been hurried off to an untimely grave, in scraping together for these gentlemen, father and son, the enormous sum of upwards of forty millions of dollars, time will not reveal.

It is this sort of work that has made England so full of cripples and paupers.

LETTER XIV.

CONDITION OF FEMALE OPERATIVES.

At a meeting recently held in Bradford,* in Yorkshire, to take into consideration the deplorable condition of 12,000 female factory operatives employed in that town, mention was made of the high moral and social condition of the female operatives of Lowell. The speakers strongly urged the necessity of a like state of things being brought about among the female operatives of England. However desirable this object may be, however praiseworthy the motives and exertions of those benevolent individuals by whom the movement has been commenced, still I feel convinced, and that conviction is founded upon a life of practical knowledge, that the object sought is beyond *their* power to obtain. Much good, it is true,

* In the history of Bradford, lately published by John James, it is stated that *five sixths* of the persons employed in the factories of this town and neighborhood, are females.

may be the result of their endeavors to ameliorate the condition of my unhappy countrywomen, and so far it is desirable to proceed; but the full measure of their wishes it is all but impossible to gain; at least, so long as the government of the country remains as at present.

It may be asked, *why* it is next to impossible. I will endeavor to show. Let us suppose one case, which will serve our present purpose. It is truly said, that the female operatives of Lowell are the daughters of farmers in comfortable circumstances; it is also well known, that those of Bradford are the daughters of poor people. Suppose, then, two single men, in the full vigor and prime of life, whom we will call Mr. Lowell and Mr. Bradford; the one being in New England, the other in Old England; but in every other respect, similarly circumstanced in life, should at one and the same time take it into their heads to marry. In doing so, we will suppose that they are actuated by the purest principles, and fully determined to do every thing in their power to make their respective families comfortable and happy. Let us also suppose, that after a certain time, each of the brides bring their happy husbands a daughter, and after other periods of time, a second, and a third. What effect will this repeated addition have upon the two families? Upon that of Mr. Lowell the effect will be trifling, so far as financial matters go. The increase of family will probably have stimulated him to increased exertion and economy, and with the great facilities, always at hand in New England, he may have become, by the time his eldest daughter arrives at the age of four years, "a farmer in comfortable circumstances."

Upon that of Mr. Bradford the effect will be very different, for he will not only have to provide increased food, clothing, lodging, &c., but for extra *taxation*, which every child will bring upon him, *from the hour of its birth*. So that by the time Mr. Lowell has become a respectable farmer in comfortable circumstances, Mr. Bradford, all his exertions to the contrary notwithstanding, will have become "a very poor man."

Look there, reader, at little Miss Lowell, plump and active, full of life and vigor, neat as a new made pin, about to be introduced to the primary school. In this school she will remain several years, surrounded by all the beneficial influences that can be brought to bear upon her character, and this, too, without any expense to her parents, beyond the purchase of a few books. After she has finished at the primary, she will be taken to a second, and afterwards, if it be thought necessary, to a finishing school. We will now suppose she has arrived at the age of fifteen, and if she has made good use of her time and privileges, she will be in possession of a fund of useful knowledge, calculated to smooth her path through time and eternity.

Let us now turn to poor Miss Bradford, for her parents have become poor; consequently she must be poor, too. See how the little timid thing creeps along, as if she felt afraid to look a person in the face. One might almost think her looks said, I could eat a buttered cake if I could get it. And though neat and clean, it is easy to see that her thin, spare clothing, is far from being sufficient to protect her from the wet and cold. Let us ask her a few questions. Pray, where are you going, my little dear?

"I am going to find old Betty."

Who is old Betty?

"The woman who minds our house while father and mother go to the factory."

And has she gone and left you?

"Yes, and little sister and baby are both crying."

By this time old Betty returns, drags her into the

house, and scolds her for leaving the children. This sort of life continues till Miss B. is six years of age, when it is concluded to discharge old Betty, on the score of economy, and leave the younger children in the charge of Miss B., getting a neighbor to look in upon them occasionally.

At nine, Miss B. is taken into the factory, having obtained a certificate from a surgeon for that purpose. In this school she is surrounded from the first hour by influences of the worst description; and if she escapes contamination, it may be looked upon almost as a miracle.

At fifteen, (the age at which we left Miss Lowell,) she will have had six years of factory life, and may be said to be well skilled, not only in her daily toil, but also in much of the vice and immorality of her elder companions.

Let us now suppose that Miss Lowell takes it into her head to go into the factories, for the purpose of "providing herself with a marriage dowry."

She enters her name, and is received as a rational and accountable being, having a soul to be lost or saved. Full provision is made in every way for her temporal and spiritual welfare. She is provided with every thing she needs to make her comfortable, and receives good wages besides. It is on these, and only these conditions that she consents to go, and why? Because she need not go to work unless she and her parents are so disposed.

In a few years she realizes the object of her wishes, saves a few hundred dollars, quits the factories, and gets married.

Miss Bradford, on the other hand, does not enter the factories from choice, but *necessity;* she has no voice in making conditions, but must submit to such as are offered. She is not looked upon by her master as an intellectual being, but as an animal *machine.* No provision is made for her temporal or spiritual welfare, as these are matters which most manufacturers trouble themselves very little about.

The small pittance she receives as wages is sometimes paid in money, sometimes partly in money, and the rest in goods. With these wages she has to provide clothes, board, lodging and washing, and also pay her *taxes* to Government.

It is a well known fact that the Government of England requires 5,000,000 dollars *weekly*, in order to enable them to pay off the interest of the *National Debt*, and carry on their Government. In order to raise this enormous demand, every man, woman and child, *must* pay a part; it matters not whether they earn twenty cents, twenty shillings, or twenty dollars a week; a part of it must be given up to Government.

Taking these things into consideration, it is no wonder if Miss B. should find herself at the close of the year as poor as at the beginning. Neither is it surprising to find her under the painful necessity of following her occupation till her strength fails, and she is turned off just the same as a machine of wood and iron, to be replaced with some new comer.

It must not be supposed that my object in writing this, is to endeavor to reduce the Lowell females to a level with those of Bradford, and elsewhere. God forbid that any such base and unworthy motive should enter the mind of any man, much less one who has been such a great sufferer by the system he is describing. My object is simply to show the condition of my countrywomen, and the insufficiency of the means about to be employed by a few benevolent persons to effect a radical change in their condition. Before any permanent good can be done, it will be necessary to pay off, or in some way erase from the statute book their enormous National Debt; reduce the extravagant expenditures of the Government; and give to the working man a voice in making the laws by which he is governed. These and other reforms are needed. They are the chief cause of the evils complained of. Remove the cause, and the effects will cease.

The effects of factory labor on females are in part illustrated by the following anecdote, related to me by a respectable linen and woolen draper of Ashton-under-lyne, with whom I had the pleasure of dining.

"A poor woman came into my shop," said he, "one Saturday night in September last, (1841) for the purpose of purchasing some small article. She had a child about twelve months old in her arms, which she set upon the counter, with its back against a pile of goods, in order that she might have her hands at liberty to examine the article she wanted. The child was not noticed by the shopmen till it became troublesome; and being Saturday night, and a great many women in the shop, I asked whose child it was, but none of the women present would take to it. A thought instantly struck me that some one had been playing the trick of child-dropping with me; however, as we were busy, I ordered the child to be brought into my parlor, and laid upon the sofa, upon which you are now sitting, where it soon fell fast asleep. About an hour afterwards, a woman came into the shop in great haste, and inquired if she had left a child there. She was brought into the parlor to see if the one lying asleep on the sofa was hers. As soon as she saw it, she cried out, 'Yes; bless thee, it is thee!' She was then asked, how she came to leave it, and by what means she had discovered her loss. To which she answered, 'That while attending to the purchase she had been making, she had quite forgotten her child. That she had been through the market, and in many other shops, and had bought all the things she wanted, but never once found out her loss till she got home, and was asked by her husband where she had left the child. To which, she said, Why, the child is up stairs asleep in bed, to be sure. But, being convinced to the contrary, and that she had taken it out with her, she began to think where she had left it. There was then no alternative but going round to every shop at which she had called; and, at last, she came to the right one.' She had left the child in the same manner as people sometimes forget their umbrellas, or a paper parcel. So you may judge," said the draper to me, "what is the effect of the system of factory labor upon these poor people and their offspring!"

I was not surprised to hear this account, well knowing, as I did, that the mothers only see their infants at morning, noon and night, except they are brought to the factory to be suckled in some other part of the day; and that for the most part, the children are in the care of strangers.

LETTER XV.

VALUE OF HUMAN LIFE IN ENGLISH FACTORIES.

In the year 1769, Mr. Richard Arkwright obtained his first patent for spinning cotton yarn, and commenced manufacturing by machinery. This was the beginning of the Factory System. Innumerable examples are furnished by history to show that at this time the inhabitants of the North and Midland counties of England, were a healthy, hardy, strong, robust people.

It was to these counties that the government looked for supplies for the Army and Navy, more particularly than to any other. In 1777, eight years after the introduction of Arkwright's (so called) invention, Manchester raised a regiment of volunteers.

This fine body of men was called the Seventy-second, or Manchester Regiment; and their gallant conduct on the rock of Gibraltar, when it was attacked by the Spaniards, and defended by General Eliot, obtained for them lasting renown. On their return to England, they were received in Manchester with enthusiasm, and their colors were deposited with much ceremony in the Collegiate Church, from whence they were removed to the College, where they still remain as trophies of the gallantry of the regiment, and of the patriotic ardor of the town.

Contrast the above with a statement made by a respectable surgeon, engaged to examine men for the militia, a few years ago; that out of 200 men examined, only *four* could be said to be *well-formed men*, and these four stated, in answer to questions from the surgeon, that they *had never worked in a factory.* This difference is believed by most people to be owing to the factory system.

According to the most moderate calculations that have been made, there are, at the present time, upwards of 10,000 bad cases of decrepitude in the factory districts, each of which can be clearly traced to the factories alone. This conclusion has been arrived at by taking a census of the cripples, in a few particular towns, and comparing these accounts with the whole district.

These cases of decrepitude are of two kinds, viz: cripples made from long standing and over exertion, and those made by accidents with machinery.

With respect to cripples from over-working, many erroneous opinions are afloat, even among the work-people themselves; the chief of which is, that some particular machines are more liable to make cripples than others. To a very limited extent this may be true; but my experience leads me to suppose, that, generally speaking, deformity is occasioned simply by *standing in one position* a greater length of time than the Divine Author of our being ever intended we should do. I am strengthened in this opinion, by the history of the cripples I have met with in different parts of the country, whether they have been brought up in the woolen, worsted, flax, cotton, or *silk* mill; nine out of every ten having been compelled to work from morning to night, without being allowed to *sit down* for a minute.

Let us turn our attention, for a moment, to the formation of the lower extremities of the human frame. There is a beautiful arch of bones formed in the foot, on the middle of which the main bone of the leg is planted; in walking, the heel and ball of the great toe touch the ground. The bones in the arch of the foot are of a wedge-like form, the same as the stones which form the arch of a bridge. This bridge receives the weight of the body, and by its elastic spring, prevents any shock being felt in leaping, &c. The weight of the body being too long sustained in factory working, this wedge-like form is lost; the bones give way, fall in, and the elastic spring of the foot is forever gone; the inside of the sole of the foot touches the ground, constituting that deformity which is called the *splay foot.* The ligaments of the ankle joint then give way, and the ankle falls inwards or outwards, as the case may be. The ligaments of the knee joint give way, causing what is called "*knock knee'd;*" or, where the leg is bent outwards, it constitutes that deformity called "*bow-legged.*" After the ligaments have given way, then the bones also bend, but not so much in the middle as at the extremities. This bending of the bones of the lower extremities is sometimes so striking, that occasionally six, or even twelve inches of height are lost in consequence; which may be proved in this manner. A man of correct proportions will, in general, be about the same height as the length of the arms from tip to tip of the long fingers, when extended. I have frequently seen cases of factory deformities, in which the length of the arms thus, was six inches more than the altitude of the body; in my own case, this difference is eight inches. I have observed, that, so far from the ratio of these cripples being in proportion to the weight of the work to be done, it is directly the contrary; the woolen, which is the heaviest employment, furnishing the fewest cripples, and the silk, which is the lightest of all, the greatest number.

A person going through a silk mill, and viewing the operations of the various branches of the manufacture, would suppose that no human beings could be deformed and crippled by such light, clean, and beautiful work; consisting of little more than knotting threads of silk, clipping the edges of ribbon, and other things, which seem to a casual observer, more suitable for a lady's parlor than a factory. But when we look more narrowly into the matter, we find causes for the awful effects of factory labor in silk mills. It may seem a very nice thing for a child or young person to be placed near a frame and have nothing to do but knot the threads of silk as they break; but, if we take into consideration that they are to remain close by that frame for twelve hours per day, and never sit down, our astonishment at the great number of silk mill cripples will vanish. Let us suppose that these young persons *had nothing to do whatever*, but were compelled to walk over a space of ground four yards long, and one yard wide, for twelve hours per day, without having leave to sit down, or rest themselves in any way, except by leaning their knees against a rail which runs along; and this duty to be performed, day after day, and year after year, the consequences, I venture to affirm, would be the same in both cases. I have also observed, that where *seats are provided*, and *extra hands kept*, so as to give the children time to rest occasionally, (as in the worsted mill of Messrs. Wood and Walker, of Bradford,) there are *no cripples made.*

In order to make myself acquainted with the number of cripples in Macclesfield, where silk manufacture is carried on to a very great extent, we had a census taken; and in this town, with a population of 24,000, 197 bad cases were found.

Deformities and diseases of the spine are a very common consequence of working in factories. I have never seen any instances of deformities of the arms, because these limbs have not to sustain the weight of the body. But even the arms share in the general weakness and debility arising from factory labor.

One evil arising from the bending and curving of the limbs, is the state of the blood vessels; for if the bones go wrong, the blood vessels must go wrong also. Nature has provided a beautiful contrivance for propelling the blood to every part of the human frame. This is done in a well-formed person with perfect ease, without any appearance of difficulty whatever. But it is not so with factory cripples. The blood lodges, as it were, in crannies and corners, and the apparatus for forcing it along, instead of being stronger, as in their case required, is weaker, in consequence of the weak state of the body. Hence we find that friction, with hair gloves, in many cases is absolutely necessary.

Females suffer greatly in after life, especially in the all-important operation, arising from the malformation of the bones of the pelvis, while standing at the frames when young.

Let us now turn to the second class of factory cripples, viz: those made so by machinery.

Accidents by machinery arise from three causes, viz: from cleaning the machinery while in motion, from the carelessness of the manufacturer in not having the machinery properly guarded, and from the carelessness of the work-people in passing and re-passing the machines. Little children, whose intellects are not sufficiently advanced to enable them to form a proper estimate of the dangers by which they are surrounded, show their tempers, have their quarrels, and push each other about, when almost in immediate contact with the most dreadful kinds of machinery; accidents of a very shocking description often occur from this cause; in addition to this, the young children are allowed to clean the machinery, actually while it is in motion; and consequently the fingers, hands and arms, are frequently destroyed in a moment. Upright and horizontal shafts, if unprotected, cause great destruction to life and limb, especially to females, whose flowing skirts get wound round while revolving at the rate of 100 to 200 times a minute. Death is frequently instantaneous.

One class of accidents arises from the *shuttle* in power-loom weaving. In large rooms where there may be upwards of 1000 shuttles flying to and fro at one time, the accidents from this cause are numerous. The shuttles are tipped with steel, and travel with great velocity, and if anything turns them out of their proper course, they in many cases pounce right upon the head of the opposite weaver, and not unfrequently turns the eye completely out of its socket on to the cheek. I have known many people who have lost one eye from this cause; one young man of my acquaintance is quite blind, having lost one eye by the shuttle, and the other by sympathy.

In order to make this matter a little more clear, let us

suppose a loom weaving a piece of common shirting. The warp, or longitudinal threads, are divided in two equal numbers, or in other words, that all the *odd* threads, counting from the side of the warp, viz : the first, third, fifth, &c., move *up* and *down* together. So, likewise, all the *even* threads, viz : the second, fourth, sixth, &c. Now, if we suppose that by a movement of the loom the *odd* threads are made to *ascend*, and the *even* threads to *descend*, they will form a sufficient space between the rows of threads for the shuttle to pass through, and leave behind it a thread of weft. The next movement of the loom will be to knock close up the weft left by the shuttle, and reverse the order of the threads of the warp, or cause the *odd* threads to *descend*, and the *even* ones to *ascend*. This movement is repeated in a power-loom from 100 to 130 times per minute; being as quick as the eye can follow the shuttle. Now, should one of the threads of the warp break, (as is frequently the case) while the loom is in full operation, the shuttle will most probably trail the broken thread across the warp, which will thus prevent the threads of the warp passing each other freely. The shuttle is thus checked on its journey, and as it is going at a railway pace, it flies out, and strikes any object that may be in its way, with a force which it would be difficult to ascertain. However, some idea may be formed of its momentum by taking into account the picks which it makes per minute, which, I before observed, are from 100 to 130; thus travelling at the rate of nine miles per hour, and making between 7000 and 8000 turnings on the road, from side to side.

One young woman who had lost an eye by the shuttle, deplored her loss very much; she was on the point of marriage when the accident took place. The loss of her eye disfigured her countenance so much, that her intended husband altered his mind. Thus was this poor girl deprived of her eye and her husband, by the *breaking of a thread of cotton yarn.*

With respect to the number of accidents by machinery, it is difficult to speak with certainty. In looking over the Reports of the Manchester Royal Infirmary for 1839, I find entered on the books, 3496 cases of accidents for that year. Of these, 2760 were out-patients, and the rest in-patients. In the same institution, in 1840, there were 3749, of which 3018 were out-patients.

It is not to be understood that all these were mill accidents; but a great many of them are recorded as such, and with respect to others, we are left in the dark. In these two years there were fifty-seven cases of amputation of legs, arms, hands and feet, in this institution.

From the Records of the Leeds General Infirmary for 1840, it appears there were received into that institution 261 cases of mill accidents, of these eleven cases required amputation.

Thus it appears that in 1840 there were on the average about five accidents a week, showing a very large amount of human misfortune, resulting from the want of precautionary measures with regard to the machinery at which the people are employed. How much greater the actual amount is, cannot be ascertained; for it must be remembered that this is a return from only one public institution, where there are several open for the reception of like accidents, independently of the private houses to which many of the sufferers apply!

Mr. Charles Trimmer, assistant inspector of factories, speaking of accidents, says, "I have taken some pains in collecting, for the last three years, from the books of the Stockport Infirmary, the number of factory accidents. The number of accidents from March, 1837, to March, 1838, in Stockport, was 120; from 1838 to 1839, 134; from March, 1839, to 1840, 86; out of which 36 were owing to their being caught whilst cleaning the machinery, the machinery being in motion at the time."

"In the Report of the Stockport Infirmary for 1839, (says Mr. Trimmer,) there is the following passage: 'The Committee cannot conclude their Report without stating a fact which was painfully impressed on their minds during the last year. They refer to the manner in which accidents generally occur in our cotton mills. Almost all the accidents that have come under the notice of the Committee, have happened in consequence of the cleaning of the machinery while it is in motion. It is earnestly hoped that the owners and managers of our manufactories will adopt effectual means for the discontinuance of so dangerous a practice.' The practice, (adds Mr. Trimmer) *has not been discontinued;* because, in the following year, when the cotton trade was very bad, there were thirty-six accidents in Stockport, owing to cleaning machinery while it was in motion." He adds, "that of 340 cases, he only knows of TWO in which the manufacturers have made any reparation or compensation to the injured party!"

I have selected these three institutions, to show the ratio of mill accidents treated in them, but it must be borne in mind that there are a great number of other institutions in the factory districts, and that, therefore, the cases here mentioned give but a very imperfect idea of the whole amount of suffering from accidents by machinery.

The following distressing factory accidents came under my personal notice, in a single journey, made a few years since. Two young men were *killed on the spot;* two other young men, and one young woman, *died in a few days,* from injuries received; one woman lost her *left arm,* another her *right arm;* one man lost *his leg,* another *his hand;* and many lost *two* and *three fingers* each.

Such statements as have from time to time come before the public respecting the factories of England, could not be circulated but upon undeniable authority; and even then, many are inclined to doubt their accuracy. A few years ago the stories in circulation were so shocking to the feelings, that men were employed for the purpose of proving their truth or falsehood.

In this way two gentlemen of unblemished character and reputation, viz: P. Ashton, M. D., and John Graham, Surgeon, undertook to examine the work people, one by one, employed in six factories of Stockport. Their report was afterwards laid before a committee of the House of Lords, and by that committee received, accepted and printed. This report is a valuable one, as showing the state and conditions of the people employed in these six factories. The factories were taken as a fair sample of all in the town. From this report, I have condensed with great labor, the following particulars.

The following table will show the number of persons employed, with the age at which they commenced working.

Age at which they began to work in the Factories.									*Employed.*		
4 years.	5 years.	6 years.	7 years.	8 years.	9 years.	10 years.	10 to 20.	Above 20.	Males.	Females.	Total.
4	35	96	147	143	112	102	151	33	429	394	823

The average age of these operatives at the time the examination took place, was eighteen years. The average time worked in factories was nine years and seventeen days each. In the previous twelve months, 182 males, and 204 females, had been off work in consequence of sickness; and the average duration of sickness was about four weeks and a half each.

Highest temperature in the mills 85 degrees. Lowest, ditto, 52. Mean, ditto, 65, 75.

The following is a tabular view of their condition.

Complaint.	*Males.*	*Females.*
Healthy	89	87
Sickly and delicate	142	172
Troubled with a cough	83	73
" " scrofula	15	12
Rheumatic affection	6	1
Bowel complaint	1	1
Difficulty of breathing	30	18
Asthma	5	1
Consumption	—	1
Pains in the head	7	18
" " back	1	—
" " breast	7	4
" " legs	1	3
Swelled legs	—	2
" ankle-joints	17	23
" knee-joints	5	6
Both knees turned in	15	2
Right knee turned in	13	15
" " " out	1	—
Left knee turned in	1	2
" " " out	1	—
Lame of both legs	—	1
Stunted in growth	39	21
Bad eyes	3	1
Swelled neck glands	—	4
Lost one arm by machinery	1	1
" a thumb by do	1	—
Lame arm by do	1	—
" hand by do	1	—
" hip by do	1	—
" leg by do	—	1
Crooked thigh	1	—
Curved legs	1	—

Hernia	1	—
Distorted Spine	—	1
Absent through sickness	3	1

These statements need no comments.

No sooner are they worked up in this way, and rendered unable to earn their living, than they are cast off, their places being supplied by new comers.

Knowing these facts, who can wonder at there being 10,000 cripples in the factory districts?

There is no provision made by the manufacturers for the support of these unfortunate persons, after being rendered useless. Had they sustained their injuries while fighting for their country, they might have looked forward to Chelsea or Greenwich Hospital; but in vain we look for such asylums for the mutilated factory cripples. There are no such institutions throughout the length and breadth of the land. The Union workhouse and the grave, are the only asylums for such cases.

Thus we behold in a Christian country, a land which boasts of being the glory and admiration of the world, thousands of human beings, mutilated and crippled, emaciated, ruined in health, their spirits broken, their minds and reasoning powers toppling from their seats, and many of them catching, like drowning men, at straws, to save themselves from what would be a happy release from their miserable situation; crying out with Job—"Wherefore is light given to him that is in misery, and life to the bitter in soul?" Contrast this with what man was intended to be. We are told that man was made "in the image of God;" that God "saw his substance yet being imperfect," and that in "His book all our members are written;" that he was made "a little lower than the angels," and "crowned with glory and honor," and placed in this lower world "to have dominion over the works of his hands." If, as we are told, all our members are written in His book, what an awful reckoning will some of these manufacturers have to meet! How will they be able to account for the lives, and limbs, which they have heedlessly, if not wantonly sacrificed?

LETTER XVI.

STATISTICAL FACTS—INCREASE OF MACHINERY—DITTO OF INDIVIDUAL LABOR—AND EARLY SUPERANNUATION OF OPERATIVES.

Spinning. Prior to the year 1767, spinning was performed by hand, every spinner tended *one* spindle; but after the discovery of Hargreaves, and the subsequent improvements of Arkwright, the spinning frame increased as follows. First it contained 12 spindles, then 24, 48, 144, 324, 648, and lastly 1028. After this, the system of coupling 2, 3, 4, and even 5 pairs of spinning mules was introduced. In 1841, a friend of mine was working five pair of spinning mules in a factory in Manchester, containing in all 3360 spindles. He was thus doing the work of 5 men, and breaking down his own constitution at the same time. His wages were about 27 shillings per week, ($6,48) which is about as much as each of the 5 men could earn on a single pair of mules in 1829. Another friend was spinning at the same time in Bolton, on 3 pair of frames containing 2400 spindles. It is said that working these frames will break the strongest constitution in six years.

In Manchester there were in 1829, 2650 spinners,

working 1,229,204 spindles; in 1841, 1037 spinners worked 1,431,619.

In Bolton in 1835, there were 30 cotton factories at work, in which were 601,226 spindles; giving employment to 798 spinners, and 2527 piecers. In 1841, of 40 factories, 38 were working, in which were 751,555 spindles, employing 737 spinners and 2457 piecers.

In addition to the increase of spindles, there is also a great advantage gained in the increase of speed. In 1817, a machine for spinning cotton yarn, called a throstle, with twelve dozen spindles, would spin one hank (containing 840 yards of cotton thread) per spindle, per day, which was considered a fair day's work. In 1841, the same sort of machine, worked by the same number of hands, and in which are eighteen dozen spindles, will spin four hanks per spindle, per day, of the same description of yarn.

According to McCullock, there were in the united kingdom in 1834, upwards of 9,000,000 of spindles employed in the cotton manufacture.

Weaving. In 1784, at a public table in Matlock, in Derbyshire, the conversation turned upon the subject of the newly invented machines for spinning cotton yarn. It was observed by some of the company present, that if this new mode of spinning by machinery should be generally adopted, so much more yarn would be manufactured than our own weavers could work up, that the consequence would be a considerable export to the continent, where it might be woven into cloth so cheaply as to injure the trade in England. Dr. Cartwright being one of the company, replied to this observation, that the only remedy for such an evil would be to apply the power of machinery to the art of weaving as well as to that of spinning, by contriving looms to work up the yarn as fast as it was produced by the spindle. Some gentlemen from Manchester who were present, and who, it may be presumed, were better acquainted with the subject of discussion, would not admit of the possibility of such a contrivance, on account of the variety of movements required in the operation of weaving. Dr. Cartwright, who, if he ever had seen weaving by hand, had certainly paid no particular attention to the process by which it was formed, maintained that there was no real impossibility in applying power to any part of the most complicated machine (producing as an instance, the automaton chess-player); and that whatever variety of movements the art of weaving might require, he did not doubt but that the skilful application of mechanism might produce them. The discussion having proceeded to some length, it made so strong an impression on Dr. Cartwright's mind, that immediately on his return home he set about endeavoring to construct a machine that should justify the proposition he had advanced, of the practicability of weaving by machinery.

In 1787, three years after the above conversation took place, the Rev. gentleman established a spinning and weaving factory in Doncaster. This factory contained 20 power looms, 10 for weaving muslin, 8 for common cottons, 1 for sail-cloth, and 1 for colored check; the machinery was all worked by a *bull*, and not till 1789 by steam power.

Imperfect as Dr. Cartwright's early machinery may seem to be, we find several eminent individuals complimenting him upon the beauty of its productions, and among others, Dr. Thurlow, bishop of Durham, to whose lady the inventor had presented a piece of muslinette. The bishop thus writes in Oct. 1787:—"Mrs. Thurlow has determined to put herself into a dress made out of the piece of muslinette you were so good to present her, and which, for its novelty, and being the first fruits of your labor and art, she prizes beyond the richest productions of the east."

Having made application to the legislature, Dr. Cartwright, in 1809, received a parliamentary grant of £10,000, ($50,000) "for the good service he had rendered the public by his invention of weaving."

There is no knowing to what extent this invention may extend. I remember being very much surprised, when for the first time I walked over a weaving room in one of the cotton mills of Lancashire. The room contained 1058 power looms, all busily at work, excepting a few that were undergoing repairs. These looms give employment to 521 weavers, chiefly young men and women; 22 overlookers, 18 twisters, 6 winders, 2 drawers, and 2 heald-pickers; in all, 571 persons. These looms are capable of weaving 3900 cuts per week, of 69 hours; each cut or piece averaging 46 yards long, 36 inches wide, 52 picks, or threads of weft to the inch. I was taken through a store-room adjoining, nearly filled with goods. I was told that there were about 300,000 pieces in that store ready for the market.

In 1835, there were 116,801 power looms in Great Britain; in 1841, the number was about 130,000, attended by about 50,000 persons, chiefly women and children.

In 1814, a *hand-loom weaver* could weave 2 pieces of nine-eighths shirting per week, each 24 yards long and 100 shoots of weft to the inch.

In 1823, a *steam-loom weaver* could weave 7 pieces; in 1826, 12 to 15 pieces; in 1833, with an assistant, 18 pieces similar to the above, in the same space of time.

By returns made to Parliament, dated 28th of March, 1836, and 20th of February, 1839, it appears that in three counties only, viz., Lancashire, Cheshire, and the West Riding of Yorkshire, the steam and water power to turn machinery in factories was increased, in three years, from 45,836 to 65,395 horse power, or 42½ per cent. While the number of hands employed in those factories had only increased in the same period, from 242,099 to 292,179; that is, only 20½ per cent."

The following table will show the increase of trade, and the decrease of wages.

	Lbs. of cotton consumed.	Hand-weaver's wages.
1797,	23,000,000,	26s. 8d a week.
1804,	61,364,158,	20s. 0d "
1811,	90,309,668,	14s. 7d "
1818,	162,122,705,	8s. 9d "
1825,	202,546,869,	6s. 4d "
1835,	333,043,464,	5s. 6d "
1840,	460,000,000,	5s. 6d "

The above is taken from McCulloch, Porter, and Fielden.

In the year 1839, there were employed in the factories of England, 419,590 persons of all ages and both sexes. Of these, 242,296 were females, and 177,294 males. Of this number, 112,119 females, and 80,768 males, were under 18 years of age.

According to the demonstrations of one of the ablest mathematicians in Europe, who was engaged to go down to the factory districts for the purpose of making calculations on the spot, a great proportion of these people *walk*, in attending the machinery, from 20 to 30 miles per day, in addition to their various other duties.

A very large proportion of these people die early in life; but should they live to 40, they are rendered incapable of doing their duty in the mills.

In 1832, in Lanark, of 1600 persons employed in the mills, only 10 were above 45 years of age.

In 1839, in 42 mills in Stockport and Manchester, in

which 22,094 hands were employed, only 143!! were above 45 years of age.

In the large factory towns, a great proportion of the poor persons selling oranges, nuts, &c. in the streets, are worn out factory workers;—thus in Manchester, of 37 hawkers of nuts, &c., 32 were factory hands; of 28 hawkers of boiled sheep's feet, 22 were factory hands. Hard return for a life of toil and industry.

LETTER XVII.

WAGES—STRIKES AND TURN-OUTS FOR WAGES—MEANS USED BY THE MANUFACTURERS TO PREVENT THEM—TEN HOUR SYSTEM—CONCLUSION.

Wages. Smith, in his "Wealth of Nations," has the following important truth. "What are the common wages of labor, depends everywhere upon the contract usually made between two parties, whose interests are by no means the same; the workman desires to get as much, and the master to give as little as possible. The former is disposed to combine in order to raise, the latter in order to lower the wages of labor."

In no country has this truth been exemplified in a more striking manner, than in England. *Trades Unions* have been introduced, on the one hand, for the purpose of protecting the wages of the laborers; and *combinations* of the manufacturers on the other, to reduce wages to the lowest possible amount.

The following table will shew the amount of money lost in the following strikes.

Cotton Spinners of Manchester, 1810,	£224,000
" " " 1826,	200,000
" " " Since	176,000
Spinners of Preston	74,313
Town of Preston	107,096
Glasgow Cotton Spinners	47,600
Loss to the City of Glasgow	200,000
Loss to the County of Lanarkshire	500,000
Strike in the Potteries	50,000
Leeds Mechanics' Strike, Twelve Months	187,000
Wool Combers of Bradford, Ten Months	400,000
Colliers	50,000
	£2,216,009

About *eleven millions* of dollars; and if all the other strikes and turn-outs were taken into account, it would swell the amount to over 20 millions of dollars, spent in a vain attempt to protect the wages of labor. Whilst the English capitalist can make use of the law to crush the producer, the producer can never make use of the law to protect himself; witness the case of the Dorchester laborers, and Glasgow cotton spinners, so well known; and, also, the case of the Stockport weavers, in 1840. The Stockport cotton masters offered a reduction of 2s. in the 12s.; and, availing themselves of the power given to them by the combination laws, caused several of the turn-out weavers to be arrested for *conspiracy*. For what? one count of the indictment was, "That they conspired to raise their wages." On this they were tried, convicted, and imprisoned. We might also instance the case of the six poor girls in the flax mills at Dundee, described on pages 91 and 92.

Let us see now what steps have been taken by the manufacturers to *reduce* wages.

In the appendix (C) to the first annual report of the Poor Law Commissioners, the reader will find a letter from Edmund Ashworth, Esq., a wealthy cotton manufacturer, from which I make the following extracts:

Turton, near Bolton, 9th of 6th month, 1834.

Respected friend, E. Chadwick :—I take the liberty of forwarding for consideration, a few observations on the new poor law bill; the leading principles of which I most cordially approve. * * * I would not venture to suggest an opinion to you, who have already so ample a store of evidence, were it not that I feel so much *the vast importance of the subject.*

Full employment in every department was never more easy to be found than now, consequently wages have advanced in most operative employments, and particularly so in the least skilful. * * * This bespeaks a scarcity of laborers here; at the same time, great complaints are made of surplus population in the agricultural counties, whilst here our deficiency is made up by a vast influx from Ireland, of ignorant, discontented, and turbulent people. * * * * It is often the practice here, if a mill owner is short of work-people, to apply to overseers of the poor and workhouses, for families supported by the parish; of late this has not always been attended with success; —— ——, (these manufacturers are supposed to be the Messrs. Gregg,) who are extensive cotton spinners and manufacturers, having two establishments in Cheshire and three in Lancashire, have, like ourselves, been in this practice many years; and being this spring short of hands at most of their establishments, sent a person who had occasionally gone out for them during a period of 20 years, to seek families in the neighboring parishes; but this year he could not find an overseer in all the county of Cheshire, who was willing to allow (compel) a family to leave his parish.

I am most anxious that every facility be given to the removal of laborers from one county to another, according to the demand for labor; this would have a tendency to equalize (reduce) wages, as well as prevent in degree, some of the turn-outs, &c., which have been of late so prevalent.

The following extracts are from a letter addressed by Robert Hyde Gregg, Esq., another cotton manufacturer, to Edwin Chadwick, Esq., Secretary to the Poor Law Commission.

Manchester, 17th September, 1834.

I have for some time thought of addressing you on the same matter as my friend Ashworth did, some time ago ; viz., the propriety of opening a communication between our (strange to say) underpeopled districts and the southern overpeopled ones.

It is at this moment a most important suggestion, and deserves to be put into immediate operation.

At this moment our machinery in one mill has been standing for twelve months for want of hands. In another mill we cannot start our new machinery for the same want.

The suggestion I would make is this; that some official channel of communication should be opened in two or three of our large towns with your office, to which the overcharged parishes might transmit lists of their families. Manufacturers short of laborers, or starting new concerns, might look over the lists and select, as they might require (for the variety of our wants is great,) large families or small ones, young children or grown up, men or widows, or orphans, &c.

If this could be done, I doubt not, in a short time, as the thing became known and tried, we should gradually absorb a considerable number of the surplus laborers of the south.

The English laborers are much preferred to the Irish, and justly so. We cannot do with refuse population, and insubordinate paupers. Hard working men, or widows with families, would be in demand.

This gentleman concludes his letter by expressing a fear that should there be any increase in the demand for laborers, it will increase the *trades unions*, *drunkenness and high wages*.

A third extract I shall make from a letter of Henry Ashworth, Esq., a brother to the first mentioned; he is also a manufacturer.

Turton, near Bolton, 2d month 13th, 1835.

Respected friend, E. Chadwick:—I have received thy letter, and the published account of the destitute condition of 32 poor families. I wish they were here, or as many of them as are willing to work.

I may safely state that there is in this neighborhood a greater scarcity of work-people than I have ever known, and this fact was never more universally acknowledged.

I am happy to say, that I never heard of any sort of privation among them, (the laborers) except what has been occasioned by their strikes on account of wages.

A great many other letters from manufacturers might be quoted, did our limits allow. They all express the same sentiments, viz., that it is impossible for too many hands to be sent—they are wanted, and must be had.

The Poor Law Commissioners, after deliberating upon the above suggestions, issued the following circular.

"The Poor Law Commissioners are desirous of facilitating it, (the removal of families) *by every means in their power,* and they therefore wish to acquaint you, (the manufacturers) that in case of your wanting the *labor of even a single family,* the commissioners proffer the use of the means at their disposal, for facilitating the supply of your wants in this respect."

Thus the reader will see that when the manufacturers could not induce a *single overseer* to give up their poor to be worked in the factories, they had recourse to the Poor Law Commissioners, who here offer them all the means at their disposal to enable them to effect their purpose.

It would occupy too much of our space were we to follow in detail, this scheme of the manufacturers to reduce the wages of labor. It will be sufficient for our purpose to say, that an office was opened in Manchester, under the superintendence of a Government Agent, R. M. Muggeridge, Esq.; that a printed circular invoice was sent to every parish where there was a surplus of poor families, which invoice was filled up by the parish officer and returned to Manchester. These invoices, containing the age, sex, &c. of each family, were kept in the office by Mr. Muggeridge, and exhibited to the manufacturers, who affixed their name and address to the invoice of such families as would suit them; after which, these precious documents were again transmitted to the parishes, with a request that they might be sent down to the manufacturers forthwith. The term of their engagement was for three years.

In this manner 10,000 persons, mostly agricultural laborers, were removed from their homes and the scenes of their childhood, and in a great majority of cases, entirely against their will, (being poor they were supposed to have no will,) to be immured in the cotton mills at Lancashire, &c.

Let us now turn our eyes to the closing scene in this tragedy. From the Report of Mr. Muggeridge, dated July, 1837, we learn that—"The condition of the laborers who were induced to migrate under your sanction, would, I am sure, at any time, be a subject of great interest and anxiety to you (the Commissioners); and at the present period, this interest and anxiety cannot but be increased by the altered circumstances in which the district is placed, compared with those which led to the introduction of migration."

"Nearly every one of the possible causes anticipated, as likely to lead to the ultimate ill success or defeat of this branch of the Commission, has been suddenly and unexpectedly realized. * * * The entire trade of the district was all at once paralyzed; distrust and want of confidence suspended for a season almost all commercial operations; the demand for additional labor ceased; large numbers of the *native* work-people were temporarily thrown out of occupation; and the extended preparations which had been made for increasing the means of employment, were deferred or abandoned."

I will not harass the feelings of my readers by going into a detail of the great hardships and sufferings endured by this class of poor people, in parting from friends and home, in their passage of 200 or 300 miles in open boats or wagons, in being herded together and lying on straw in the ware-houses of the manufacturers on their arrival, or arising from small pox, fevers, accidents by machinery, or a want of faith on the part of the manufacturers afterwards. They were such as would scarcely be credited as having taken place in England, a land always boasting of her humanity, and such as England herself would not have tolerated in any other country, without an effort to suppress or ameliorate them.

The reader may imagine the extent to which this evil had been carried, from the following remarks made by Sir Robert Peel, in Parliament, on Tuesday evening, January 27th, 1846.

"I will now," said he, "direct the attention of the house to a law which has been greatly complained of by the agricultural interest; I mean the present law of settlement. It happens under the present law, that a large portion of the population of rural districts are induced to move into the manufacturing; and it happens frequently, that the power, the labor, the best part of a man's life, who so removes, are consumed in that manufacturing district, and thus all the advantages of his strength and good conduct, and industry, are derived by the manufacturing districts during the period of his residence. A revulsion then takes place in trade; and what course is taken with respect to the man who moved there in more prosperous times? The man with his wife and family are sent back to the rural districts; and the individual who spent the best part of his life as a manufacturing operative, is returned to the place from whence he came; returned unfit for an agricultural operative. Under these circumstances the man is sent, *against his will*, to a new home, at a period when all his communications with that district have been interrupted, and with no means of earning an honest livelihood, a proceeding which must shock the feelings of every man who witnesses it.

"At present, when a man, situated as the individual I have described, begins to *fail* in a manufacturing district—perhaps from having undergone extreme labor, or from sickness or accident, and apprehension begins to be entertained that he may become a permanent incumbrance on the parish, means are promptly taken for an early removal of that man. Again, immediately on the death of a laboring man in a manufacturing district, of which he happens not to be a native, his widow and children can be removed to the parish in which they had a previous settlement.

"We propose, that after a man has labored for a period of five years in a district, his settlement shall not be in the place where he had originally a settlement; but in the district to which his industry and labor had been given during that five years. That there shall not only be no power to remove that man, but that there shall be no power to remove his wife or his children, legitimate or illegitimate, under sixteen years of age."

In this manner the manufacturers gained their point with regard to breaking up trades unions, and reducing the wages.

The ten hour bill. I perceive by an account in the newspapers, that the English Parliament have at length agreed on a law for limiting the hours of labor in factories to ten per day; to go into operation in May, 1848. This *ought* to be a wise and good law, considering the length of time it has been before Parliament, (namely,

32 years,) the amount of talent employed in framing it, and the enormous sums of money expended in fostering it in its embryo state, till it became a thing of life. It may be useful and profitable to some, if we take a hasty retrospective glance at this bill in its various stages, from its first framing up to the present time.

In the year 1802, the late Sir Robert Peel, member of the House of Commons, and cotton manufacturer, procured an act (42 Geo. 3, c. 73) to regulate the labor of apprentice children worked in factories. At this time, the children in factories were mostly apprentices, obtained from parishes in London and other large towns.

The Apprentice Act, naturally, but gradually wore out the custom of taking apprentices, for as the masters would work the long hours, they now had recourse to the children of parents on the spot. This movement was hastened by the application of Watt's steam engine, which enabled the manufacturer to erect his factories in large towns, instead of, as formerly, on the banks of streams.

The evils sought to be remedied by the above act still continuing, Sir Robert again came before Parliament on the 13th of June, 1815, and proposed a ten hour bill, for regulating the hours of labor in *cotton* factories; and a committee of the House of Commons agreed to *ten and a half*. This was the first movement on record that I am acquainted with, for a ten hour bill.

In the following year, 1816, the same, or a similar bill was referred to a select committee of the House of Commons. The evidence taken before this committee established many important facts concerning the hardships and cruelties endured by the children employed in factories, both apprentices and those who were not apprentices.

The evidence of John Moss, overseer of Backbarrow

Mill, is to this effect,—that the apprentice act was constantly set at naught. The witness did not even know of it. The children in the mill were almost all apprentices from London parishes; they were worked from 5 in the morning till 8 at night, all the year round, with only one hour for the two meals; in making up lost time, they frequently worked from 5 till 10 at night; and invariably they worked *from* 6 *on the Sunday morning till* 12, in cleaning the machinery for the week. In speaking of the consequent fatigue, the evidence is this:

"Did the children sit or stand at work?"

Stand.

"The whole of their time?"

Yes.

"Were there any seats in the mill?"

None.

"Were they much fatigued at night?"

Yes, some of them were very much fatigued.

"Where did they sleep?"

In the apprentice house.

"Did you inspect their beds?"

Yes, every night.

"For what purpose?"

Because there were always some of them missing; some might be run away, others I have sometimes found asleep in the mill.

"Upon the mill floor?"

Yes.

"Did the children frequently lie down upon the mill floor at night when their work was over, and fall asleep before their supper?"

I have found them frequently upon the mill floor asleep, after the time for bed.

This, and other similar evidence, was quite sufficient to justify Sir Robert in applying for a ten hour bill. The bill, however, did not pass then, but was suffered to slumber till July, 1819; but this (59 Geo. 3, c. 66) did not apply to any but cotton factories.

After the passing of this Act, there were four others to amend, to alter, or to render valid this one; but these were all repealed by the 1 and 2 William, 4. c. 39, commonly called Sir John Hobhouse's Act. The principal provision of this act is one which makes it unlawful to work in a factory, any child who is under 18 years of age, for more than 69 hours in a week; but this act also is confined to cotton factories.

In 1832, the late Mr. Sadler made great efforts in favor of the factory children. He brought a bill into Parliament to limit the hours of labor for *all* under 18 years of age, to 58 hours in the week; and the provisions of this bill were to extend to woollen, flax and silk, as well as cotton mills. On moving the second reading, on the 13th of March, he was met by strong opposition, and a cry for investigation. He acceded to a committee being called out, of which he became the chairman. Before this committee, a mass of evidence was adduced which surprised and astonished all who heard it for the first time Physicians, surgeons, manufacturers, overlookers and operatives, both male and female, were examined and the two large volumes of evidence thus collected, will stand as a lasting reproach to the country to the latest time. Mr. Sadler, although he labored almost day and night, till his health was greatly impaired, did not obtain his much desired object.

It is worthy of observation, that about this time, the Parliament passed an act to abolish slavery in the English colonies, and not only the *name*, but the *essence* of slavery; for, in that act, it has taken care to provide that no negro shall work more hours in the week than 45, which is no more than 7 1-2 in a day. Now, if this act of humanity was necessary, which no one will doubt, how much more necessary the 58 hour act for the offspring of Englishmen!

On the meeting of the first reformed Parliament, Mr. Sadler not being a member, Lord Ashley was prompted to take the question in hand. He was beaten in the House of Commons, on the motion of Mr. Patten, because the ministers thought they could protect the children without interfering with the adults.

It would much exceed our present limits to examine in detail, all the various movements of the friends and foes of this bill for the last fifteen years. Motions and counter motions in the House, reports and counter reports out of it, have been constantly before the public. Sometimes one party have had the lead, and sometimes the other. One now proposing to "lay the axe at the root and cut it (the factory system) down;" another begging that the country would consider first what sort of a tree it was, and look at the fruit it had produced.

It is matter of congratulation, however, that the bill has been suffered to have an existence, if for nothing more than to test its merits in removing the evils of which the factory people complain. Of its success I have strong reasons to doubt.

If there is any merit in the ten hour system, the benevolent portion of the manufacturers will come in for their share. We find the first man to move in it was a manufacturer; and since then, many petitions have been sent to the House, and the ministers, for the bill, by manufacturers. Lord John Russell presented petitions signed by nearly 500 master manufacturers, employing 140,000 hands, for the bill.

It is also curious to observe that Sir Robert Peel *opposed the measure*, although he is the son of a cotton spinner, and a son of the very first man who brought forward the motion in Parliament in 1815!

Among the most prominent advocates for the bill, we find the name of Gould, Oastler, Fielden, Sadler, Rev. G. S. Bull, Lord Ashley, Wood, Walker and Brotherton; some of these men have spent large sums of money for their favorite object.

Let us now consider a few of the most prominent points to be effected by this measure.

It is expected, in the first place, that a great benefit will be gained by the laborers, in point of health. This, however, will be much greater in some instances than in others. A comparative view of the several branches of manufacture into which "the factory system" divides itself, will enable us to see this more clearly.

The material to be manufactured may be summed up under the following names: cotton, flax, tow, wool, worsted and silk. The three first of these are *vegetable*, the last three, *animal* substances. It is much more healthy to work the animal than the vegetable material.

Cotton, it is well known, is extremely light in weight, and its fibre is short and buoyant. It is, therefore, from the nature of its staple to be expected, that there will be a considerable portion of the "flyings," or waste of the cotton, held in suspension in the atmosphere of a cotton mill. These fine particles of cotton and dust being taken into the throat, lungs and stomach, at each inspiration of air by the laborers, engender a morbid irritability of those members, and ultimately induce a chronic disease called *Gastraglia*. A fixed and incurable asthma, consumption, or premature death, is frequently the result. It is also well known that cotton, but especially the finer sort, will be best manufactured at or about the same temperature as that in which it is grown; hence the artificial atmosphere of English cotton mills is extremely detrimental to health.

In flax and tow mills, the same results follow from much the same causes, viz : the breathing of the fine particles of flax and dust held in suspension in the atmosphere of these mills. There is one portion of fine flax mills which is exempt from these pernicious ingredients; but the female laborers in these rooms have to be constantly with their hands in hot water, surrounded by steam, which makes these rooms equally, or perhaps more unhealthy than the others. This is in consequence of each thread of fine linen yarn having to pass through hot water, in the act of spinning. This makes the fibres of flax move more freely among each other, and enables the manufacturer to produce much finer thread than he could possibly do without the artificial heat applied in this way.

The woolen and worsted mills are free from all the peculiarities of the foregoing branches of trade. In woolen mills, it is true, there is an effluvia arising from the oil, which is profusely used, in addition to the natural animal grease, or yolk of the wool; and likewise from the dye, in which, in some branches, the wool is prepared before its processes commence in the mill. But the worsted, which consists of the longest fibres of the fleece, must first be worked perfectly clean, and rendered as free as possible from the natural animal grease, and other impurities. This I think a material point of distinction in favor of the healthiness of the worsted trade.

The short fibres which are disengaged, and fly off as waste in the various processes of the woolen and worsted manufacture, are not of the like injurious nature as cotton "flyings," as they are too heavy to be held in suspension in the atmosphere, and accordingly, instead of floating in it, they fall to the ground.

Of the *silk* manufacture, I should think, from the nature

of the material, and its great value, one might safely state it to be as cleanly and healthy an occupation as can be followed in a factory. Of this, however, my experience does not warrant me in speaking with certainty.

From these observations it will be clear, that to people working in factories *all the year round*, subject to these pernicious influences, it will be less dangerous to health to remain at work ten hours per day, than twelve or thirteen.

The second point of consideration for working classes is *wages;* for it is, beyond all doubt, well understood, that a proportionate reduction of wages will be generally adopted when the factory bill becomes law. The working classes, who, though often deluded upon points of speculative opinion, on account of their partial information, are generally shrewd and sound in their practical views, will see the cause of the change, and the alleged reason for it, and will ultimately, if not immediately, become convinced that a reduction of their wages, without a corresponding reduction in their taxation, and other grievous burdens, will not relieve them from the oppressive evils of which they complain.

Hence it follows, as a matter of course, that the great debt of the nation must in some way be cancelled, or removed, before the people will feel themselves much benefited by the ten hour bill.

I know there are some who will ask, what has *debt*, or *taxation* to do with the hours of labor? and what good can come of mixing up these questions with the factories bill? To such I would answer, that these evils are the *very root of the matter;* and the adequate causes of the effects in question; and so long as they continue to exist it matters not whether men work ten hours a day, or twelve; for these will hang like a mill-stone about their necks, weighing them down lower and lower—and not only laboring men, but all the productive classes, till the employers approach nearer and more near every year, to the condition of the employed.

The next question is, How are the people to employ their newly acquired leisure? If they resort to the ale-house to spend the evening, or to theatrical, and other places of amusement, or establishments for gambling, they will have gained but little by the change. The national system of education may do much for the rising generation, but will not affect the great body of factory operatives at present employed.

Here then is a great field for the exercise of the time and talent of the benevolent portion of the community, for the members of churches and temperance societies.

It would be presumption to suppose that the modern system of manufacturing industry, which concentrates hundreds of individuals in the same apartments, can be divested of its inevitably contingent evils and disadvantages by an act of Parliament; and that the dense manufacturing population of England can be reduced to a primitive state of simplicity and innocence of manners by a bill which leaves them in the very midst of all that is opposite and counteracting.

There are other important points connected with the ten hour factory bill—such as the commercial bearing of the question, the currency, the population, and a few others—but these I will leave to abler pens than mine, my chief object being to show the effect it will probably have on the working classes.

The bill, taken by itself, is not of so much importance as at first sight one might be led to suppose; but as a link in the great chain of improvements already introduced, and which *must* be introduced into the social system of England, it is of great importance.

CONCLUSION.

In taking a hasty glance at the factory system, two things are evident to the most casual observer.

First, that the wealth of the country, and especially of certain districts, has increased greatly within the last forty years;—Second, that *the race of Englishmen is dwindling down, and degenerating under the effects of the unremitting labor*, and the insufficient and unwholesome food that their country's laws allows them to enjoy.

The creation of wealth, is, in many instances, the multiplication of happiness; but there are circumstances which render such creation anything but a blessing. When the wealth created is generally distributed, it is an unmixed good, and always associated with the progress of civilization. It is never generally distributed, *but when the capitalists employed are the many, and not the few.*

That the wealth of the country has greatly increased, is not a matter of doubt. This is not only an acknowledged truth, but it is the boast of the millocracy. It is at the head of all their demands—"Mark our wealth—our importance to the conntry." This is the language in which their demands are urged. It is then agreed that they are opulent—that they have vast estates—that they are not only able to buy, but have actually bought up a portion of the aristocracy. We have now one point fully established, viz: that the millocracy abound in riches, which have been regularly accumulating for a series of years. This fact suggests a simple question. Have the artisans employed in those branches of trade participated in the benefits of these riches? Has their condition been progressively improving? To avoid misconception, we will descend to particulars. Have the wages of the artisans gradually increased within the last quarter of a century? No! They have declined. Do the artisans live better? No! Many die, because they are not able to live. Do they clothe better? No! Thousands are confined at home on the Sabbath day for want of clothes. Have their dwellings become more comfortable? No! In some of the largest manufacturing towns, one seventh of the population live in cellars. Has the necessity for infant labor diminished with the accumulation of these riches? No! It has increased in successive years to such an alarming extent, that the legislature has been compelled to interfere, to arrest the sacrifice of the miserable infants. Are the artisans better educated? No! Two thirds cannot write their names, and the proportion that can read and write becomes less in each successive generation. Are they more moral? No! Crime has increased in a greater ratio than population. Has the average duration of life increased? No! It has greatly diminished. In one of the largest manufacturing towns in the kingdom, in 1822, 1 in 44 of the population died; but in 1837, 1 in 24½. The millocracy intrude on public consideration their importance as a wealthy class; they state in figures the immense capital which a small number can command; they also admit the misery, wretchedness, immorality, and degradation of the artisans.

One of the favorite axioms of the millocracy is, "that capital owes no allegiance to soil," and, consequently, if not petted, will take wings and depart. The man was shrewd who asked what posterity had done for him; and the masses with equal shrewdness, may ask the millocracy, "what has this capital done for us?" We perceive it in abundance in your hands, but there is none in ours. You are legislators as well as manufacturers—evidence that capital has served your end, but pray how are we benefited, either by your capital or legislation? You are elegantly clothed, and sumptuously fed, whilst we are in rags, and struggling against difficulties to support existence. Luxuries and comfort await you in your mansion. A poisonous atmosphere and cheerless poverty await us in our miserable abode. You are a new order, influential from your wealth. We are an old order, who have become beggared and exhausted in giving birth to you. You are the patrons of charity. We are the recipients of it. Contemporaneous with the extension of your magnificent factories, is the establishment of a new system of parochial relief; new modes of punishment, and enlarged conveniences for the reception of felons. Capital may have given the soft pillow to your head, and a flowery path to your footsteps; to us it has made an easy transition from the factory to the prison and the poor-house! The doctrine, "that capital owes no allegiance to the soil," is one of the finest illustrations that ever presented itself to the mind, of the gross principle of selfishness pervading the thoughts, feelings, and visions of the millocracy. Capital has increased beyond the means of profitable employment—it has increased to overflowing; and contemporaneous with this excess is the augmentation of poverty, wretchedness, and crime, in the humble instruments of its creation. Does capital owe no allegiance to those who have produced it? Is it, after having been wrung out of their exertions, to depart to happier climes? Can no portion be spared to alleviate the misery which its production has occasioned? The doctrine, that capital owes no allegiance to the soil, is certainly perfectly new. It never, in any country or at any time, in its most shadowy or indistinct form, suggested itself to the mind. Important truths are generally got at piecemeal. One amount of discovery leads to another, until the truths, whole or in majestic fragments, break upon the understanding. This truth, however, sent no shadow before it, nor had it any labor-pains, though labor had produced it.

In ancient times, when capital abounded, the fine arts sprang into vigorous existence. The statuary made the marble all but breathe. The painter with his exquisite art, touched the canvass into life. The philosopher taught his spiritual and humanizing doctrines. The poet, fresh from nature and glowing with divine conceptions, awoke impassioned eloquence in the listening crowd. Capital produced these effects! Capital was encouragement, and owned an allegiance to every thing that was grand, refined, or elevated in nature.

In more modern times, when capital flowed abundantly into the lap of the Italian States, the poet, the sculptor, the painter, again felt its invigorating spirit, and threw life, beauty, and imagination over the gross realities of existence. Religion, the instinct of our nature, arose, adorned in all the captivating luxury of genius. But where is now this instinctive feeling—this love of the Deity, that prompts to noble deeds? We ask not for the form which shows where it is not. Why does the admiration of man live upon the achievements of the past? Is human nature stunted in its growth? Is it dwindling into insignificance for want of encouragement? What! remove the capital—leave the most glorious of all fields uncultivated—the boundless faculties of the human soul! Is there no duty involved in this capital? You acknowledge none. Capital, when its creation has been a blessing to its producers, has no tendency to escape. It feels the attractive influence of the soil, and remains with it. It is as loath to quit it, as fragrance the flower around which it lingers.

I have now brought to a close my remarks upon the English factory system; a system which is utterly at variance with the perfect law of God, and which contains within itself the means of its own destruction. It is now about seventy-seven years since it first commenced under Sir Richard Arkwright, and I believe there are few persons at the present day so sanguine as to believe that it will live till it is 100 years old. A new and a better system has commenced in America.

To the people in this country, therefore, the remarks contained in this book are of the utmost importance. There can be no doubt that America is destined to become the first manufacturing country in the earth. If through the good influence of the valuable institutions of this highly favored land, its inhabitants are led to choose the good and avoid the evil, as I am happy to perceive is the case in New England, the factory system, so far from being a *curse*, will be one of its greatest blessings. That such may be the case, may God in his infinite mercy grant.

A

VOICE FROM THE FACTORIES.

I.

When fallen man from Paradise was driven
Forth to a world of labor, death and care;
Still, of his native Eden, bounteous Heaven
Resolved one brief memorial to spare,
And gave his offspring an imperfect share
Of that lost happiness, amid decay;
Making their first *approach* to life seem fair,
And giving, for the Eden past away,
CHILDHOOD, the weary life's long happy holyday.

II.

Sacred to heavenly peace, those years remain !
And when with clouds their dawn is overcast,
Unnatural seem the sorrow and the pain
(Which rosy joy flies forth to banish fast,
Because that season's sadness may not last.)
Light is their grief ! a word of fondness cheers
The unhaunted heart; the shadow glideth past;
Unknown to them the weight of boding fears,
And soft as dew on flowers, their bright, ungrieving tears.

III.

See the Stage-Wonder (taught to earn its bread
By the exertion of an infant skill,)
Forsake the wholesome slumbers of its bed,
And mime, obedient to the public will.
Where is the heart so cold that does not thrill
With a vexatious sympathy, to see
That child prepare to play its part, and still
With stimulated airs of gaiety
Rise to the dangerous rope, and bend the supple knee?

IV.

Painted and spangled, trembling there it stands,
Glances below for friend or father's face,
Then lifts its small round arms and feeble hands,
With the taught movements of an artist's grace:
Leaves its uncertain gilded resting place—
Springs lightly as the elastic cord gives way—
And runs along with scarce perceptible pace—
Like a bright bird upon a waving spray,
Fluttering and sinking still, whene'er the branches play.

V.

Now watch! a joyless and distorted smile
Its innocent lips assume; (the dancer's leer!)
Conquering its terror for a little while:
Then lets the TRUTH OF INFANCY appear,
And with a stare of numbed and childish fear
Looks sadly towards the audience come to gaze
On the unwonted skill which costs so dear,
While still the applauding crowd, with pleased amaze,
Ring through its dizzy ears unwelcome shouts of praise.

VI.

What is it makes us feel relieved to see
That hapless little dancer reach the ground;
With its whole spirit's elasticity
Thrown into one glad, safe, triumphant bound?
Why are we sad, when, as it gazes round
At that wide sea of paint, and gauze, and plumes,
(Once more awake to sense, and sight, and sound,)
The nature of its age it re-assumes,
And one spontaneous smile at length its face illumes?

VII.

Because we feel, for Childhood's years and strength,
Unnatural and hard the task hath been;—
Because our sickened souls revolt at length,
And ask what infant innocence may mean,
Thus toiling through the artificial scene;—
Because at that word, Childhood, start to birth
All dreams of hope and happiness serene—
All thoughts of innocent joy that visit earth—
Prayer—slumber—fondness—smiles—and hours of rosy mirth.

VIII.

And therefore when we hear the shrill faint cries
Which mark the wanderings of the little sweep;
Or when, with glittering teeth and sunny eyes,
The boy-Italian's voice, so soft and deep,
Asks alms for his poor marmoset asleep;
They fill our hearts with pitying regret,
Those little vagrants doomed so soon to weep—
As though a term of joy for all was set,
And that *their* share of Life's long suffering was not yet.

IX.

Ever a toiling *child* doth make us sad;
'Tis an unnatural and mournful sight,
Because we feel their smiles should be so glad,
Because we know their eyes should be so bright.
What is it, then, when, tasked beyond their might,
They labor all day long for others' gain,—
Nay, trespass on the still and pleasant night,
While uncompleted hours of toil remain?
Poor little FACTORY SLAVES—for YOU these lines complain!

X.

Beyond all sorrow which the wanderer knows,
Is that these little pent-up wretches feel;
Where the air thick, and close, and stagnant grows,
And the low whirring of the incessant wheel
Dizzies the head, and makes the senses reel:
There, shut forever from the gladdening sky,
Vice premature and Care's corroding seal
Stamp on each sallow cheek their hateful dye,
Line the smooth open brow, and sink the saddened eye.

XI.

For them the fervid summer only brings
A double curse of stifling withering heat;
For them no flowers spring up, no wild bird sings,
No moss-grown walks refresh their weary feet;—
No river's murmuring sound;—no wood-walk, sweet
With many a flower the learned slight and pass;—
Nor meadow, with pale cowslips thickly set
Amid the soft leaves of its tufted grass,—
Lure *them* a childish stock of treasures to amass.

XII.

Have we forgotten our own infancy,
That joys so simple are to them denied?—
Our boyhood's hopes—our wanderings far and free,
Where yellow gorse-bush left the common wide
And open to the breeze?—The active pride
Which made each obstacle a pleasure seem;
When, rashly glad, all danger we defied,
Dashed through the brook by twilight's fading gleam,
Or scorned the tottering plank, and leapt the narrow stream?

XIII.

In lieu of this,—from short and bitter night,
Sullen and sad the infant laborer creeps;
He joys not in the glow of morning's light,
But with an idle yearning stands and weeps,
Envying the babe that in its cradle sleeps:
And ever as he slowly journeys on,
His listless tongue unbidden silence keeps;
His fellow laborers (playmates hath he none)
Walk by, as sad as he, nor hail the morning sun.

XIV.

Mark the result. Unnaturally debarred
All nature's fresh and innocent delights,
While yet each germing energy strives hard,
And pristine good with pristine evil fights;
When every passing dream the heart excites,
And makes even *guarded* virtue insecure;
Untaught, unchecked, they yield as vice invites;
With all around them cramped, confined, impure,
Fast spreads the moral plague which nothing new shall cure.

XV.

Yes, this reproach is added; (infamous
In realms which own a Christian monarch's sway!)
Not suffering *only* is their portion, thus
Compelled to toil their youthful lives away:
Excessive labor works the SOUL's decay—
Quenches the intellectual light within—
Crushes with iron weight the mind's free play—
Steals from us LEISURE purer thoughts to win—
And leaves us sunk and lost in dull and native sin.

XVI.

Yet in the British Senate men rise up
(The freeborn and the fathers of our land!)
And while these drink the dregs of Sorrow's cup,
Deny the sufferings of the pining band.
With nice-drawn calculations at command,
They prove—rebut—explain—and reason long;
Proud of each shallow argument they stand,
And prostitute their utmost powers of tongue
Feebly to justify this great and glaring wrong.

XVII.

So rose, with such a plausible defence
Of the unalienable RIGHT OF GAIN,
Those who against Truth's brightest eloquence
Upheld the cause of torture and of pain:
And fear of Property's Decrease made vain,
For years, the hope of Christian Charity
To lift the curse from SLAVERY's dark domain,
And send across the wide Atlantic sea
The watchword of brave men—the thrilling shout, "BE
FREE!"

XVIII.

What is to be a slave? Is't not to spend
A life bowed down beneath a grinding ill?—
To labor on to serve another's end,—
To give up leisure, health, and strength, and skill—
And give up each of these *against your will?*
Hark to the angry answer:—" Theirs is not
A life of slavery; if they labor,—still
We *pay* their toil. Free service is their lot;
And what their labor yields, by us is fairly got."

XIX.

Oh, Men! profane not Freedom! Are they free
Who toil until the body's strength gives way?
Who may not set a term for Liberty,
Who have no time for food, or rest, or play,
But struggle through the long, unwelcome day
Without the leisure to be good or glad?
Such is their service—call it what you may.
Poor little creatures, overtasked and sad,
Your Slavery hath no name,—yet is its Curse as bad!

XX.

Again an answer. " 'Tis their parents' choice.
By *some* employ the poor man's child must earn
Its daily bread; and infants have no voice
In what the allotted task shall be: they learn
What answers best, or suits the parents' turn."
Mournful reply! Do not your hearts inquire
Who tempts the parents' penury? They yearn
Toward their offspring with a strong desire,
But those who starve *will* sell, even what they most require.

XXI.

We grant their class must labor—young and old;
We grant the child the needy parents' tool:
But still our hearts a better plan behold;
No bright Utopia of some dreaming fool,
But rationally just, and good by rule.
Not against Toil, but Toil's Excess we pray,
(Else were we nursed in Folly's simplest school)
That so our country's hardy children may
Learn not to loathe, but bless, the well apportioned day.

XXII.

One more reply! The *last* reply—the great
Answer to all that sense or feeling shows,
To which all others are subordinate:—
"The Masters of the Factories must lose
By the abridgment of these infant woes.
Show us the remedy which shall combine
Our equal gain with their increased repose—
Which shall not make our trading class repine,
But to the proffered boon its strong effects confine."

XXIII.

Oh! shall it then be said that Tyrant acts
Are those which cause our country's looms to thrive?
That Merchant England's prosperous trade exacts
This bitter sacrifice, e'er she derive
That profit due, for which the feeble strive?
Is her commercial avarice so keen,
That in her busy, multitudinous hive
Hundreds must die like insects, scarcely seen,
While the thick-thronged survivors work where they have been?

XXIV.

Forbid it, Spirit of the glorious Past,
Which gained our Isle the surname of 'The Free,'
And made our shores a refuge at the last
To all who would not bend the servile knee,
The vainly-vanquished sons of Liberty!
Here came the injured, the opprest,
Compelled from the Oppressor's face to flee—
And found a home of shelter and of rest
In the warm, generous heart that beat in England's breast.

XXV.

Here came the Slave, who straightway burst his chain,
And knew that none could ever bind him more;
Here came the melancholy sons of Spain;
And here, more buoyant, Gaul's illustrious poor
Waited the same bright day that shone before.
Here rests the Enthusiast Pole! and views afar
With dreaming hope, from this protecting shore,
The trembling rays of Liberty's pale star
Shine forth in vain to light the too-unequal war!

XXVI.

And shall REPROACH cling darkly to the name
Which every memory so much endears?
Shall *we*, too, tyrannise,—and tardy Fame
Revoke the glory of our former years,
And stain Britannia's flag with children's tears?
So shall the mercy of the English throne
Become a by-word in the Nations' ears,
As one who pitying heard the stranger's groan,
But to these nearer woes was cold and deaf as stone.

XXVII.

Are there not changes made which grind the Poor?
Are there not losses every day sustained,—
Deep grievances, which make the spirit sore?
And what the answer, when *these* have complained?
"For crying evils there hath been ordained
The REMEDY OF CHANGE; to obey its call
Some individual loss must be disdained,
And pass as unavoidable and small,
Weighed with the broad result of general good to all."

XXVIII.

Oh! such an evil *now* doth cry aloud!
And CHANGE should be by generous hearts begun,
Though slower gain attend the prosperous crowd,
Lessening the fortunes for their children won.
Why should it grieve a father, that his son
Plain competence must moderately bless?
That he must trade, even as his sire has done,
Not born to independent idleness,
Though honestly above all probable distress?

XXIX.

Rejoice! Thou hast not left enough of gold
From the lined heavy ledger, to entice
His drunken hand, irresolutely bold,
To squander it in haggard haunts of vice:—
The hollow rattling of the uncertain dice
Eats not the portion which thy love bestowed;—
Unable to afford that PLEASURE's price,
Far off he slumbers in his calm abode,
And leaves the Idle Rich to follow Ruin's road

XXX.

Happy his lot! For him there shall not be
The cold temptation given by vacant time;
Leaving his young and uncurbed spirit free
To wander thro' the feverish paths of crime!
For *him* the Sabbath bell's returning chime
Not vainly ushers in God's day of rest;
No night of riot clouds the morning's prime:
Alert and glad, not languid and opprest,
He wakes, and with calm soul is the Creator blest.

XXXI.

Ye save for children! Fathers, is there not
A plaintive magic in the name of child,
Which makes you feel compassion for *their* lot
On whom Prosperity hath never smiled?
When with your OWN an hour hath been beguiled
(For whom you hoard the still increasing store),
Surely, against the face of Pity mild,
Heart-hardening Custom vainly bars the door,
For that less favored race—THE CHILDREN OF THE POOR.

XXXII.

"The happy homes of England!"—they have been
A source of triumph, and a theme for song;
And surely if there be a hope serene
And beautiful, which may to Earth belong,
'T is when (shut out the world's associate throng,
And closed the busy day's fatiguing hum),
Still waited for with expectation strong,
Welcomed with joy, and overjoyed to come,
The good man goes to seek the twilight rest of home.

XXXIII.

There sits his gentle Wife, who with him knelt
Long years ago at God's pure altar-place;
Still beautiful,—though all that she hath felt
Hath calmed the glory of her radiant face,
And given her brow a holier, softer grace.
Mother of SOULS IMMORTAL, she doth feel
A glow from Heaven her earthly love replace;
Prayer to her lip more often now doth steal,
And meditative hope her serious eyes reveal.

XXXIV.

Fondly familiar is the look she gives
As he returns, who forth so lately went,—
For they *together* pass their happy lives;
And many a tranquil evening have they spent
Since, blushing, ignorantly innocent,
She vowed, with downcast eyes and changeful hue,
To love Him only. Love fulfilled, hath lent
Its deep repose; and when he meets her view
Her soft look only says,—"I trust—and I am true."

XXXV.

Scattered like flowers, the rosy children play—
Or round her chair a busy crowd they press;
But, at the FATHER's coming, start away,
With playful struggle for his loved caress,
And jealous of the one he first may bless.
To each, a welcoming word is fondly said;
He bends and kisses some; lifts up the less;
Admires the little cheek, so round and red,
Or smooths with tender hand the curled and shining head.

XXXVI.

Oh! let us pause, and gaze upon them now.
Is there not one—beloved and lovely boy!
With Mirth's bright seal upon his open brow,
And sweet, fond eyes, brimful of love and joy?
He, whom no measure of delight can cloy,
The daring and the darling of the set;
He who, though pleased with every passing toy,
Thoughtless and buoyant to excess, could yet
Never a gentle word or kindly deed forget?

XXXVII.

And one, more fragile than the rest,—for whom
As for the weak bird in a crowded nest—
Are needed all the fostering care of home,
And the soft comfort of the brooding breast:
One, who hath oft the couch of sickness prest!
On whom the Mother looks, as it goes by,
With tenderness intense, and fear suppress,
While the soft patience of her anxious eye
Blends with "God's will be done,"—"God grant thou may'st not die!"

XXXVIII.

And is there not the elder of the band?
She with the gentle smile and smooth, bright hair,
Waiting, some paces back,—content to stand
Till these of Love's caresses have their share;
Knowing how soon his fond paternal care
Shall seek his violet in her shady nook,—
Patient she stands—demure, and brightly fair—
Copying the meekness of her Mother's look,
And clasping in her hand the favorite story-book.

XXXIX.

Wake, dreamer!—Choose;—to labor Life away,
Which of these little precious ones shall go
(Debarred of summer-light and cheerful play)
To that receptacle for dreary woe,
The Factory Mill?—Shall He, in whom the glow
Of Life shines bright, whose free limbs' vigorous tread
Warns us how much of beauty that we know
Would fade, when *he* became dispirited,
And pined with sickened heart, and bowed his fainting head?

XL.

Or shall the little quiet one, whose voice
So rarely mingles in their sounds of glee,
Whose life can bid no living thing rejoice,
But rather is a long anxiety;—
Shall he go forth to toil? and keep the free
Frank boy, whose merry shouts and restless grace
Would leave all eyes that used his face to see,
Wistfully gazing towards that vacant space
Which makes their fireside seem a lone and dreary place?

XLI.

Or, sparing these, send Her whose simplest words
Have power to charm—whose warbled, childish song,
Fluent and clear, and bird-like, strikes the chords
Of sympathy among the listening throng,—
Whose spirits light, and steps that dance along,
Instinctive modesty and grace restrain:
The fair young innocent who knows no wrong,—
Whose slender wrists scarce hold the silken skein
Which the glad Mother winds; shall *She* endure this pain?

XLII.

Away! The thought—the *thought* alone brings tears!
THEY labor—*they*, the darlings of our lives!
The flowers and sunbeams of our fleeting years;
From whom alone our happiness derives
A lasting strength, which every shock survives;
The green young trees beneath whose arching boughs
(When failing Energy no longer strives,)
Our wearied age shall find a cool repose;—
THEY toil in torture!—No—the painful picture close.

XLIII.

Ye shudder,—nor behold the vision more!
Oh, Fathers! is there then one law for these,
And one for the pale children of the Poor,—
That to their agony your hearts can freeze;
Deny their pain, their toil, their slow disease;
And deem with false complaining they encroach
Upon your time and thought? Is yours the Ease
Which misery vainly struggles to approach,
Whirling unthinking by, in Luxury's gilded coach?

XLIV.

Examine and decide. Watch through his day
One of these little ones. The sun hath shone
An hour, and by the ruddy morning's ray,
The last and least, he saunters on alone.
See where, still pausing on the threshold stone,
He stands, as loth to lose the bracing wind;
With wistful wandering glances backward thrown
On all the light and glory left behind,
And sighs to think that HE must darkly be confined!

XLV.

Enter with him. The stranger who surveys
The little natives of that dreary place
(Where squalid suffering meets his shrinking gaze,)
Used to the glory of a young child's face,
Its changeful light, its colored sparkling grace,
(Gleams of Heaven's sunshine on our shadowed
earth !)
Starts at each visage wan, and bold, and base,
Whose smiles have neither innocence nor mirth,—
And comprehends the Sin original from birth.

XLVI.

There the pale Orphan, whose unequal strength
Loathes the incessant toil it *must* pursue,
Pines for the cool sweet evening's twilight length,
The sunny play-hour, and the morning's dew :
Worn with its cheerless life's monotonous hue,
Bowed down, and faint, and stupified it stands;
Each half-seen object reeling in its view—
While its hot, trembling, languid little hands
Mechanically heed the Task-master's commands.

XLVII.

There, sounds of wailing grief and painful blows
Offend the ear, and startle it from rest;
(While the lungs gasp what air the place bestows,)
Or misery's joyless vice, the ribald jest,
Breaks the sick silence : staring at the guest
Who comes to view their labor, they beguile
The unwatched moment; whispers half suppresst
And mutterings low their faded lips defile,—
While gleams from face to face a strange and sullen smile.

XLVIII.

These then are his Companions : he, too young
To share their base and saddening merriment,
Sits by : his little head in silence hung;
His limbs cramped up; his body weakly bent;
Toiling obedient, till long hours so spent
Produce Exhaustion's slumber, dull and deep.
The Watcher's stroke,—bold—sudden—violent,—
Urges him from that lethargy of sleep,
And bids him wake to Life,—to labor and to weep !

XLIX.

But the day hath its End. Forth then he hies
With jaded, faltering step, and brow of pain;
Creeps to that shed,—his HOME,—where happy lies
The sleeping babe that cannot toil for Gain;
Where his remorseful Mother tempts in vain
With the best portion of their frugal fare :
Too sick to eat—too weary to complain—
He turns him idly from the untasted share,
Slumbering sinks down unfed, and mocks her useless care.

L.

Weeping she lifts, and lays his heavy head
(With all a woman's grieving tenderness)
On the hard surface of his narrow bed;
Bends down to give a sad unfelt caress,
And turns away;—willing her God to bless,
That, weary as he is, he need not fight
Against that long-enduring bitterness,
The VOLUNTARY LABOR of the Night,
But sweetly slumber on till day's returning light.

LI.

Vain hope ! Alas ! unable to forget
The anxious task's long, heavy agonies,
In broken sleep the victim labors yet !
Waiting the boding stroke that bids him rise,
He marks in restless fear each hour that flies—
Anticipates the unwelcome morning prime—
And murmuring feebly, with unwakened eyes,
"Mother ! Oh Mother ! is it yet THE TIME ?"
Starts at the moon's pale ray—or clock's far distant chime.

LII.

Such is *his* day and night ! Now then return
Where your OWN slumber in protected ease;
They whom no blast may pierce, no sun may burn;
The lovely, on whose cheeks the wandering breeze
Hath left the rose's hue. Ah ! not like these
Does the pale infant-laborer ask to be :
He craves no tempting food—no toys to please—
Not Idleness, but less of agony;
Not Wealth,—but comfort, rest, CONTENTED POVERTY.

LIII.

There is, among all men, in every clime,
A difference instinctive and unschooled :
God made the MIND unequal. From all time
By fierceness conquered, or by cunning fooled,
The World hath had its Rulers and its Ruled :—
Yea—uncompelled—men abdicate free choice,
Fear their own rashness, and, by thinking cooled,
Follow the counsel of some trusted voice;—
A self-elected sway, wherein their souls rejoice.

LIV.

Thus, for the most part, willing to obey,
Men rarely set Authority at naught :
Albeit a weaker or a worse than they
May hold the rule with such importance fraught;
And thus the peasant, from his cradle taught
That some must *own*, while some must *till* the land,
Rebels not—murmurs not—even in his thought.
Born to his lot, he bows to high command,
And guides the furrowing plough with a contented hand.

LV.

But, if the weight which habit renders light
Is made to gall the Serf who bends below—
The dog that watched and fawned, prepares to bite !
Too rashly strained, the cord snaps from the bow—
Too tightly curbed, the steeds their riders throw—
And so, (at first contented his fair state
Of customary servitude to know,)
Too harshly ruled, the poor man learns to hate
And curse the oppressive law that bids him serve the Great.

LVI.

Then first he asks his gloomy soul the Cause
Of his discomfort; suddenly compares—
Reflects—and with an angry spirit draws
The envious line between his lot and theirs,
Questioning the Justice of the unequal shares.
And from the gathering of this discontent,
Where there is strength, Revolt his standard rears;
Where there is weakness, evermore finds vent
The sharp annoying cry of sorrowful complaint.

LVII.

Therefore should Mercy, gentle and serene,
Sit by the Ruler's side, and share his Throne :—
Watch with unerring eye the passing scene,
And bend her ear to mark the feeblest groan;
Lest due Authority be overthrown,
And they that ruled perceive (too late confest !)
Permitted Power might still have been their own,
Had they but watched that none should be opprest—
No just complaint despised—no WRONG left unredressed.

LVIII.

Nor should we, Christians in a Christian land,
Forget who smiled on helpless infancy,
And blest them with divinely gentle hand.
" Suffer that little children come to me : "
Such were His words to whom we bow the knee !
These to our care the Saviour did commend;
And shall we HIS bequest treat carelessly,
Who yet our full protection would extend
To the lone Orphan child left by an Earthly Friend?

LIX.

No ! rather what the Inspired Law imparts
To guide our ways, and make our path more sure;
Blending with Pity (native to our hearts,)
Let us to these, who patiently endure
Neglect, and penury, and toil, secure
The innocent hopes that to their age belong;
So, honoring Him, the Merciful and Pure,
Who watches when the Oppressor's arm grows strong,
And helpeth them to right—the Weak—who suffer wrong!

ANONYMOUS.